*Spectral
and
Chemical Characterization
of
Organic Compounds*

# Spectral
## and
# Chemical Characterization
## of
# Organic Compounds

## A Laboratory Handbook

**W. J. CRIDDLE** and **G. P. ELLIS**

*Department of Chemistry,*
*University of Wales Institute of Science and Technology,*
*Cardiff*

*A Wiley—Interscience Publication*

**JOHN WILEY & SONS**
London · New York · Sydney · Toronto

*Library of Congress Cataloging in Publication Data:*

Criddle, W. J.
    Spectral and chemical characterization of organic compounds.

    'A Wiley—Interscience publication.'
    Includes index.
    1. Chemistry, Organic. 2. Spectrum analysis.
I. Ellis, Gwynn Pennant, joint author. II. Title.
QD272.S6C73    1976          547'.3          75-23296

ISBN 0 471 18767 4 (Cloth)
ISBN 0 471 01499 0 (Pbk)

Typesetting by Preface Limited, Salisbury, Wiltshire and
printed by the Pitman Press Ltd. Bath

# PREFACE

Most courses in organic chemistry require that students be familiar with the spectral properties of organic compounds and also be able to recognize compounds by virtue of their characteristic chemical reactions. By combining these two approaches and correlating the results, it is usually possible to decide on the structure of the compound and, during the course of the work, the student learns a great deal of organic chemistry in a relatively short time. It also introduces to the student the methods which are adopted in research when a compound of unknown structure is found in nature or is synthesized in the laboratory.

Characterization of organic compounds in this way requires that the student have conveniently at hand the spectral and chemical data which have been accumulated over the years and which are spread through several theoretical books. The purpose of this handbook is to bring together the information that the student is likely to need in the laboratory and to present it in the most convenient way so that the minimum of time is used in searching for relevant information. There are several good theoretical books available on spectroscopy and therefore some knowledge of them is assumed. Similarly, we assume that the basis of the chemical tests in Chapter 2 will have been obtained from lectures and books on organic chemistry.

Systematic names of organic compounds have, as far as possible, been used throughout the book. At last, there are real signs that the old trivial names are being superseded in schools and colleges. We have included many of these as alternative names. The scope of the book is widened by the inclusion of a section on the characterization of pharmaceutical compounds. This section (Chapter 6) will be of particular interest to students of pharmacy.

We wish to thank Dr. D. J. Bailey (Welsh School of Pharmacy) for his valuable assistance with Chapter 6 and Professor W. H. Hunter (Chelsea College, University of London) for useful comments. We also thank the publishers for their patience and co-operation and Mrs. P. Bevan and Mrs. J. M. Watkins for secretarial assistance.

W. J. CRIDDLE
G. P. ELLIS

# CONTENTS

# PRELIMINARY TESTS

## 1. ELEMENTAL ANALYSIS

The identification of the elements contained in an organic compound is a first and most important step in organic analysis and is best effected by using Lassaigne's test. In this test the organic compound is decomposed by fusion with sodium. The presence of the elements nitrogen, halogens, sulphur or phosphorus in the original compound is then determined by various tests on the product. If nitrogen is present in the compound, fusion with sodium converts it into sodium cyanide which may be identified by the reaction of the cyanide ion in solution. Halide, sulphide or phosphate ions in the product similarly indicate the presence of halogen, sulphur or phosphorus in the compound being analysed.

Lassaigne's test does not provide information on the presence of carbon, hydrogen or oxygen. The vast majority of organic compounds contain carbon and hydrogen and it is usually possible to identify a compound without testing specifically for oxygen although the ferrox test (described below) gives a positive test for the majority of oxygen-containing compounds.

### Lassaigne's test

CAUTION: Metallic sodium must be handled with great care as it reacts violently with water and many other compounds. It must never be allowed to come into contact with the skin and protective goggles should be worn.

*For solids.* Place a piece of metallic sodium (about 2 mm cube) in an ignition tube and heat gently until it is molten. Add the organic compound (about 200 mg) and continue heating gently until the contents of the tube are solid. Then heat more strongly and maintain at red heat for 2 min. While the tube is still red hot, plunge it into a 50 cm$^3$ beaker containing distilled water (15 cm$^3$). Boil for 3—4 min and filter the solution. Use the filtrate (A) for the tests given below.

*For liquids.* Heat a piece of metallic sodium (about 2 mm cube) in an ignition tube until the tube is one-third full of sodium vapour. Introduce the liquid (0.2 cm$^3$) dropwise into the tube using a dropping-tube. When all the liquid has been added, heat strongly for 2 min and plunge the red-hot tube into a 50 cm$^3$ beaker containing distilled water (15 cm$^3$). Boil for 3—4 min and use the filtrate A for the following tests.

### Nitrogen

To filtrate A (1 cm$^3$) add ferrous sulphate solution (10%, 1 cm$^3$) and a little 2 M sodium hydroxide solution until a heavy precipitate of ferrous hydroxide is obtained. Boil this for 2 min, cool and acidify (test with litmus paper) with 2 M sulphuric acid. A precipitate or coloration varying from deep blue to green indicates the presence of *Nitrogen* in the organic compound.

## Sulphur

To filtrate A (1 cm$^3$) add freshly prepared sodium nitroprusside solution. A pink to purple coloration indicates the presence of *Sulphur*. This coloration is sometimes transient.

## Halogens

To filtrate A (1 cm$^3$) add excess of 2 M nitric acid (test with litmus) and if nitrogen or sulphur has been detected by the above tests, boil the solution for 5 min in a beaker (fume cupboard) to remove hydrogen cyanide or hydrogen sulphide. Boiling is not necessary if nitrogen and sulphur are absent. To the cool solution add silver nitrate solution. A white or yellow precipitate indicates the presence of one or more of *Chlorine, Bromine* or *Iodine* in the organic compound. If this test is positive, the identity of the halogen will be revealed by the following tests, but before these are attempted it is advisable to carry out a blank test on the reagents alone.

If it is known that only one halogen is present it may be identified as follows: to filtrate A acidified with dilute sulphuric acid, add chloroform (1 cm$^3$) and chlorine water or 1% sodium hypochlorite solution (2 drops). Shake well and allow the chloroform layer to separate. A brown coloration in the chloroform layer indicates *Bromine*, a purple coloration *Iodine*, while no change in the colour of the chloroform indicates *Chlorine*.

If more than one halogen may be present, the following series of tests should be carried out:

(*a*) To the filtrate A (2 cm$^3$) add an excess of 2 M nitric acid followed by 5% mercuric chloride solution (1 cm$^3$) (POISON). A yellow precipitate which changes to orange or red on standing for a few minutes indicates the presence of *Iodine*. If a high concentration of iodide ions is present in the solution, the precipitate is orange or red immediately.

(*b*) To filtrate A (2 cm$^3$) add an equal volume of dichromate oxidizing mixture and boil gently for 2 min. Test the vapours produced with a filter paper dipped in freshly prepared Schiff's reagent. A purple coloration indicates the presence of *Bromine*.

(*c*) To filtrate A add an excess of 2 M nitric acid and then silver nitrate solution. Filter off the precipitate and treat it with an excess of a solution consisting of four volumes of saturated ammonium carbonate solution and one volume of ammonia solution (0.88). If a precipitate remains, filter this off and acidify the filtrate with dilute nitric acid. A white precipitate indicates the presence of *Chlorine*. It should be noted that silver bromide is slightly soluble in the solution used above. A faint precipitate obtained during this test should therefore be ignored.

(*d*) Acidify a further portion of filtrate A (2 cm$^3$) with acetic acid; bring the solution to the boil and cool. Add a drop of this solution to a piece of filter paper dipped in zirconium-alizarin solution (1% ethanolic alizarin and 0.4% aqueous zirconium nitrate) and allow the paper to dry. A red to yellow colour change indicates the presence of *Fluorine*.

## Phosphorus

Treat a further portion of filtrate A (2 cm$^3$) with concentrated nitric acid (0.5 cm$^3$) followed by a 5% solution of ammonium molybdate. Heat on a boiling-water bath for 2 minutes. A yellow precipitate indicates the presence of *Phosphorus*.

### Ferrox test for oxygen

Grind together equal weights of potassium thiocyanate and ferric alum. Place the mixture (about 100 mg) in a test tube and add the organic compound directly if a liquid or as a saturated solution in benzene or chloroform if a solid. A purple coloration in the organic layer indicates the presence of oxygen in the compound. This test is specific for oxygen only if nitrogen and sulphur are absent.

## 2. IGNITION

Place some of the organic compound (0.1 g) on a spatula or in an ignition tube and heat until ignition occurs. Remove from the flame and observe the ignition characteristics. A clear flame indicates an aliphatic compound while a smoky flame is characteristic of aromatic and some unsaturated compounds. Continue the ignition until no further change occurs; the presence of a residue shows that the original compound contains a metal atom and the residue should be examined by the standard inorganic procedures for the identification of metals. In the majority of cases, a flame test carried out on the residue, which should be acidified by addition of concentrated hydrochloric acid (1 drop), will be sufficient to identify the metal atom present. It is sometimes possible to recognize the odour of the vapours released during ignition; some compounds (for example, carbohydrates, aliphatic hydroxy acids and their salts) char readily while others (for example, benzoic acid) sublime.

## 3. COLOUR AND ODOUR

The majority of organic compounds are colourless when pure but some compounds become discoloured on standing due to the formation of small amounts of coloured impurities. If a pure compound is coloured, it will contain one or more chromophoric groups, for example, nitro, nitroso or azo group, or be a quinone or have an extended conjugated system of four or more double bonds. The nitro group on its own confers very little if any colour on a compound but if an auxochromic substituent such as a hydroxyl or amino group is also present, the very pale yellow colour is intensified. An indication of the colour of many compounds is given in the tables of melting points at the end of this book.

Some organic compounds have characteristic odours which can be used tentatively to guide us in organic analysis but since smell cannot often be described in words, the student's best approach should be to try to 'memorize' the smell of some common compounds.

## 4. DETERMINATION OF PHYSICAL CONSTANTS

### Melting point

Before a melting point is determined, the sample must be pure and free of solvent. Seal off a standard melting-point tube at one end in a Bunsen flame. Introduce the sample to a depth of about 2 mm at the sealed end of the tube (rubbing the tube with a nail file often facilitates this); attach the tube with a thin wire (or rubber band) to a short-immersion $360^\circ$ thermometer with the sample at same level as the thermometer bulb. Clamp the thermometer at a depth of about 3 cm in an oil-bath (the rubber band should be above the oil level) and then heat gently over a gauze with a micro-burner. Stir continuously while adjusting the burner to give a heating rate not greater than $5^\circ$/min. The temperature at which a meniscus forms from the molten sample is the required melting point.

Notes: (a) It is often better to determine the approximate melting point of the sample first and then redetermine the melting point more accurately on a second sample by slowing down the heating rate to $2^\circ$/min as the approximate melting point is approached.

(*b*) For samples which are thermally unstable it is preferable to determine the approximate melting point in the ordinary way and then to determine the accurate value by heating the oil-bath to within 10° of the rough value before inserting the sample; the temperature is then raised at 2°/min until the compound melts. This reduces the period during which the sample is being heated and thus minimizes thermal decomposition.

(*c*) Where an electrically heated melting point apparatus is available, it should be used according to the instructions provided with the apparatus, although paragraphs (*a*) and (*b*) are valid for any melting point determination.

## Mixed melting point

When two different compounds are mixed together and the melting point of the mixture is determined, it is found that melting begins at a temperature several degrees below that of the lower-melting pure compound. This technique of *mixed melting points* may therefore be used to determine whether or not the two samples are identical. A depression in the melting point of a sample when mixed with another indicates that the compounds are different. The correct procedure for such a determination is as follows: grind together equal weights of the known and unknown materials and introduce the mixture into a melting-point tube. Place a little of each of the two pure compounds in two other tubes and determine the melting points of all three simultaneously. If the mixture melts more than 5° below the melting point of either of the pure samples, the latter are different. If they are identical, the three samples will melt at the same temperature.

## Boiling point

### Siwoloboff's method

Take a glass tube about 5 cm long and 0.5 cm internal diameter and a standard melting-point tube and seal each at one end only. Introduce the liquid under examination (0.5 cm$^3$) into the larger tube and place in it the melting-point tube, *open end* in the liquid. Attach the tubes to a short-immersion 360° thermometer with the liquid at the same level as the thermometer bulb. Immerse the thermometer in a liquid paraffin bath to a depth of 3 cm. Heat the bath at a constant rate with continuous stirring until a steady stream of bubbles emerges from the lower end of the smaller tube. Stop heating at this point and note the temperature at which the liquid rises rapidly into the smaller tube. This is the boiling point of the liquid. If the sample is impure (for example, contains a small amount of water) Siwoloboff's method will give misleading values. It is then better to remove the impurity by fractional distillation or by thorough drying with a desiccant.

## Solubility in various solvents

The solubility of an organic compound in water, ether, 2 M hydrochloric acid and 2 M sodium hydroxide can often furnish useful information about the nature of the compound. However, the presence of more than one functional group may have such a profound effect on the solubility that it is often impossible to make deductions about the functional groups present from solubility data. For example, 1,3-dihydroxybenzene is extremely soluble in water but the introduction of a butyl group into the 4-position gives a compound which is only slightly soluble. Even positional isomers sometimes differ greatly in their solubility, for example, the solubilities of 1,2-, 1,3- and 1,4-dihydroxybenzene in water at 20° are 45, 210 and 7% respectively. The following table of solubilities should therefore be used with caution, and it is likely to be most accurate for monofunctional compounds. It gives an indication of the solubility of various types of organic compounds in ether, water, 2 M hydrochloric acid and

2 M sodium hydroxide. The compounds are grouped together according to the elements which are identified in the Lassaigne test. A + sign means that the compounds of that class have a solubility in the particular solvent in excess of 5%. A lower solubility is indicated by a − sign. A class of compounds whose members vary greatly in solubility in a particular solvent is shown as ±.

## Table of solubilities

| | Ether | Water | 2 M HCl | 2M NaOH | Comments |
|---|---|---|---|---|---|
| *Compounds containing C, H, O, Metal* | | | | | |
| Carboxylic acids | | | | | |
| aliphatic | + | +[a] | +[a] | + | [a]Insol. if $>$ 4C atoms. |
| aromatic | + | − | − | + | |
| metal salts | − | + | ± | + | |
| Phenols | | | | | |
| monohydric | + | −[b] | − | + | [b]Simple phenols are sol. |
| di- and tri-hydric | + | +[c] | ± | + | [c]1,3,5-trihydroxybenzene is insol. |
| phenoxides | − | + | ± | + | |
| Aldehydes and Ketones | | | | | |
| aliphatic | + | −[d] | −[d] | −[d] | [d]Sol. if $<$4C atoms. |
| aromatic | + | − | − | − | |
| Acetals | + | ± | −[e] | ± | [e]Hydrolysis occurs |
| Alcohols | + | +[f] | +[f] | +[f] | [f]Sol. if $<$4C atoms. |
| Carbohydrates | − | + | + | + | |
| Polyols | − | + | + | + | |
| Esters | + | −[g] | −[h] | −[h] | [g]Sol. if $<$4C atoms. [h]May hydrolyse to sol. products |
| Anhydrides | + | ±[i] | − | + | [i]Aliphatic, +; aromatic, −. |
| Lactones | + | −[j] | − | + | [j]$\gamma$-Butyrolactone is sol. |
| Quinones | + | − | − | + | |
| Ethers | + | − | − | − | |
| Hydrocarbons | + | − | − | − | |
| *Compounds containing C, H, N(O)* | | | | | |
| Amines | | | | | |
| pri. aliphatic | + | + | + | − | |
| s-aliphatic | + | −[k] | + | −[k] | [k]Sol. if $<$4C atoms. |
| t-aliphatic | + | ±[l] | + | − | [l]Variable. |
| pri. aromatic | + | − | + | − | |
| s-aromatic | + | − | + | − | |
| t-aromatic | + | − | + | − | |
| Amides | − | −[m] | − | − | [m]Sol. if $<$6C atoms. |
| N-Substituted amides | + | − | − | − | |
| Imides | − | − | − | + | |
| Ammonium salts | − | + | +[n] | + | [n]Depends on solubility of free acid. |
| Nitrocompounds | + | − | − | − | |
| Amino-acids | − | ±[o] | ±[o] | ±[o] | [o]Variable. |
| Arylhydrazines | + | − | + | − | |
| *Compounds containing C, H, S(O)* | | | | | |
| Sulphonic acids | − | + | + | + | |
| Thiols and thiophenols | + | − | − | + | |

## Table of solubilities (*cont.*)

| | Ether | Water | 2 M HCl | 2 M NaOH | Comments |
|---|---|---|---|---|---|
| *Compounds containing* C, H, P(O) | | | | | |
| Phosphate esters | + | − | − | − | |
| *Compounds containing* C, H, *Halogen* (O) | | | | | |
| Alkyl and aryl halides | + | − | − | − | |
| Acyl halides | + | +q | +q | +q | qDecomposed, alkyl compounds rapidly. |
| *Compounds containing* C, H, *Halogen*, N(O) | | | | | |
| Quaternary ammonium salts | − | + | + | + | |
| Hydrohalides of organic bases | − | + | + | + | |
| *Compounds containing* C, H, *Halogen*, S(O) | | | | | |
| Sulphonyl halides | + | − | − | − | |
| *Compounds containing* C, H, N, S(O) | | | | | |
| Thioamides | −r | −r | − | + | rSol. if <3C atoms. |
| Sulphates of organic bases | − | + | + | − | |
| Sulphonamides | + | − | − | + | |

# CHEMICAL AND SPECTROSCOPIC CHARACTERIZATION OF FUNCTIONAL GROUPS

When the elements which are present in an organic compound have been determined, it is then necessary to ascertain how these are arranged in the molecule, that is, what functional group(s) it contains. For this purpose, use is made of the chemical reactions and spectral data which are characteristic of each function.

## CHEMICAL TESTS

The tests described below are grouped according to the element(s) detected in the Lassaigne test. Compounds containing only C, H, O should be examined by the tests given in Table I. If one of the elements halogen, N, P or S is present, the tests in Tables II—V should take precedence. Where two or more of these elements are present, the initial tests should be for the composite group containing the relevant elements (Tables VI—IX); for example, if N and S have been detected, the presence of a sulphonamide group ($-SO_2 \cdot NH_2$) among others should be investigated as described in Table VIII. If these tests are negative, a search should be made for the presence of groups containing the individual elements, that is, sulphur-containing functions (Table IV) and nitrogen-containing functions (Table II).

When a functional group has been identified, it should be possible to correlate this with the results of the preliminary tests so that examination of the appropriate melting point table (p. 45) will narrow the choice to one or more possible compounds. In such cases, consideration of the structure of the likely compounds may suggest further functional group tests to differentiate between them and thus enable the compound to be identified. For example, an organic compound, m.p. 122°, containing C, H, N, O was shown to be a primary aromatic amine. Reference to Table 9 shows that it may be one of three compounds having a melting point close to 122°, namely, 3-aminophenol, 2,4-dimethyl-5-nitroaniline or 4-aminobenzophenone. These may be distinguished by the appropriate tests for the second functional group, that is, phenolic, nitro and carbonyl group respectively. The identification should then be confirmed by the preparation of one or more of the derivatives given in Table IX.

A series of tests contained within heavy horizontal lines is for related functions. In many cases, if the first test in a series is negative, the subsequent tests in that particular series may be omitted.

## INFRARED SPECTRAL DATA

### Preparation of samples

(a) Solids. Solid organic compounds are normally prepared in one of two ways. Firstly, by making a solution (~10% w/v) using an anhydrous solvent which does not interfere with the spectrum of the compound. Solvents commonly used are carbon tetrachloride, trichloromethane (chloroform), carbon disulphide and hexane. Note that it is important that the solvent does not associate in any way with the substrate. Spectra may then be obtained using standard path length cells (0.025—0.1 mm). Dual-cell operation, i.e. with solvent alone in the reference beam, is sometimes useful to minimize the effects of solvent absorption, but it should be noted that this technique will not allow absorptions of substrate to be measured where strong solvent absorption occurs.

Alternatively, the solid may be converted to a mull, i.e. ground to a paste with a non-volatile oil, e.g. Nujol (a high-boiling paraffin hydrocarbon) or Flurolube (a mixture of high-boiling fluorinated hydrocarbons). Use of these two oils allows most of the usual infrared region to be scanned. The mull may be conveniently prepared using a standard B14 cone and socket joint by taking the cone and adding Nujol (1 drop) and the sample (2—5 mg) to the ground area. The cone is inserted into the socket and the two parts twisted. Mull formation is rapid and the mull can be removed from the joints using a small spatula. The mull is placed onto a rock-salt plate, covered with a second plate and pressed into a thin film. Note that when handling rock salt plates or cells, protective gloves should be worn to prevent fogging of the rock salt surfaces.

(b) Liquids. Organic liquids may be examined either as solutions (as for solids) or as films. For volatile liquids (b.p. $< 120°$) a closed cell should be used. Less volatile liquids may be examined as described for mulls, one drop of liquid usually being sufficient.

### Interpretation of spectra

(a) Most functional groups have their characteristic absorptions (stretching frequencies) in the region $4000—1400 \text{ cm}^{-1}$. This region should be examined first when a knowledge of the elemental constituents of the unknown compound has been established.

(b) When classifying the possible functional groups according to group frequencies, it is important to keep in mind the physical state of the compound when the spectrum was obtained. Spectra of solids and liquids usually show a lowering in the group frequencies of polar groups due to hydrogen bonding. The lowering of group frequencies as a result of molecular association (intermolecular bonding) is often accompanied by a broadening of the peak. Spectra obtained in dilute solution or in the vapour phase do not show these effects unless they result from intramolecular bonding.

(c) Absorptions found in the region $1400—900 \text{ cm}^{-1}$ (fingerprint region) are normally more difficult to interpret due to the complexity of group vibrations in this region. Many vibrational (bending and stretching) modes are possible in this region and assignments should be made with caution.

(d) The region below $900 \text{ cm}^{-1}$ is mainly useful for providing information regarding aromatic substitution patterns (Table X) and for some stretching frequencies of carbon—halogen groups.

(e) Abbreviations used for infrared intensities are: s, strong; m, medium; w, weak; v, variable.

## ULTRAVIOLET SPECTRAL DATA

### Preparation of samples

Both solid and liquid compounds are usually examined as solutions in a solvent which has little or no absorption in the region 220—800 nm. The concentration of the solution will depend on the absorbance value(s) at whatever absorption maxima are observable. Solvents should be of spectroscopic quality and among those commonly used are water, methanol, ethanol, chloroform, carbon tetrachloride and hexane. Cells used are either glass or silica (for absorptions below 340 nm) and usually have a 1 cm path length. Dual-cell operation as described for infrared spectroscopy is usual.

### Interpretation of spectra

Many groups which absorb in the ultraviolet show maxima over a wide wavelength range and it is therefore difficult to be specific about any particular formation. However, high absorbance values are an indication that a conjugated system is present and may be confirmed by other methods. Ultraviolet data are usually presented by quoting the molar extinction coefficient ($\epsilon_{max}$) at the appropriate wavelength ($\lambda_{max}$). Sometimes, particularly in pharmaceutical literature, the specific extinction coefficient ($E_{1cm}^{1\%}$) is quoted and is sometimes useful for characterizing a compound in much the same way as a melting-point.

# NUCLEAR MAGNETIC RESONANCE SPECTRAL DATA

### Preparation of samples

The compound should be dissolved in a solvent which itself does not absorb. Carbon tetrachloride, carbon disulphide and deuterotrichloromethane (deuterochloroform, $CDCl_3$) are commonly used. A concentration of about 10% w/v is desirable but it is often possible to work with slightly lower concentrations by adjusting the spectrometer. It is important to use a thoroughly dry solvent and to exclude moisture and water vapour until the spectrum has been obtained. For some compounds, water, deuterium oxide or dilute hydrochloric acid may be used as solvent but a part of the spectrum will be obscured by the strong solvent absorption (except for deuterium oxide).

A small amount of tetramethylsilane should be added to the solution to serve as an internal standard but since this compound is insoluble in aqueous solutions, it should be replaced by sodium 3-(trimethylsilyl)propanesulphonate.

### Interpretation of spectra

The following characteristics should be considered when interpreting the spectrum of a hydrocarbon or any organic compound containing C—H bonds:

(a) The number of protons which give rise to each signal may be deduced by measuring the fall in the integrator trace for each signal. If the molecular formula is not known, it may only be possible to determine the ratio of the number of protons in the various signals.

(b) The chemical shift of the signal should be noted and compared with that of hydrogens in known environments, for example, as listed in Tables XI and XII.

(c) The multiplicity of the signal provides information about the number of hydrogen atoms which are attached to adjacent atoms. This valuable feature applies mostly to aliphatic groups where first-order spin-spin coupling occurs (i.e. the ($n$+1) rule is valid) but may also provide information on the orientation of hydrogens on an aromatic ring.

(d) Replacement of a hydrogen by deuterium on treating the compound with $D_2O$ is accompanied by the disappearance of a signal and indicates the presence in the compound of a hydrogen attached to oxygen, nitrogen or sulphur.

**Table I.** Functional Groups Containing C, H, O

| Chemical test | Observation | Functional group | Spectral data |
|---|---|---|---|
| 1. Dissolve the organic compound (0.5 g) in water or aqueous ethanol (5 cm³) and add M sodium hydroxide solution (1 drop). Test the solution with BDH '1014' Indicator (2 or 3 drops). | Green coloration | Carboxyl (RCOOH) Phenol (ArOH) Enol (RC=CHR')<br>$\quad\quad\ \mid$<br>$\quad\quad$ OH | Carboxyl. I.R. $3000-2500$ cm$^{-1}$, R = Ar and Alk, $\nu$ O–H (broad and complex); $1720-1710$ cm$^{-1}$, R = Alk, $\nu$ C=O(m) (lowered by $\alpha\beta$-unsaturation); $1700-1680$ cm$^{-1}$, R = Ar. $\nu$ C=O(m); $900-860$ cm$^{-1}$, (m, broad). N.M.R. $\delta$ 10.0–13.0, unaffected by dilution |
| a. Prepare a saturated solution of the compound in 50% aqueous ethanol and add an equal volume of 5% sodium bicarbonate solution | Carbon dioxide evolved | Carboxyl | Phenol. I.R. $3620-3610$ cm$^{-1}$. $\nu$ O–H (w, very sharp – no hydrogen bonding); $3150-3050$ cm$^{-1}$, $\nu$ O–H (s, broad – normal absorption for hydrogen-bonded phenols); $1410-1310$ cm$^{-1}$, $\delta$ O–H (m, broad); $1230$ cm$^{-1}$ $\nu$ C–O (s, broad); $\sim 650$ cm$^{-1}$, $\gamma$ O–H (variable). |
| b. (i) Dissolve in water and add aqueous neutral iron(III) chloride (1 or 2 drops) (ii) Dissolve in methanol and add 5% methanolic anhydrous iron(III) chloride | Wide range of colorations (See Tables) | Phenol or enol | N.M.R. $\delta$ 4.5–6.8, varies with concentration. If hydrogen bonded, $\delta$ 8.0–13.0. Enol I.R. Similar to phenols, but with differences due to presence of alkenic C=C and (usually) absence of characteristic aromatic absorption (see hydrocarbon absorption below). |
| c. If Test 1b is positive, prepare a cold solution of mercury(I) nitrate in 2 M nitric acid and add the organic compound | Immediate grey precipitate of mercury | Enol | N.M.R. $\delta$ 14.0–17.0 for O–H stabilized by hydrogen bonding N.B. The O–H $\rightarrow$ O–D exchange is facile in D$_2$O for all O–H groups |
| 2. Treat the organic compound with 2,4-dinitrophenylhydrazine in either 5 M hydrochloric acid (for water-soluble compounds) or ethanolic phosphoric acid. N.B. Acetals are hydrolysed in acid solutions to aldehydes and will give a positive test with 2,4-dinitrophenyl-hydrazine. Silver mirror tests will be negative unless the acetal is first hydrolysed. | Yellow to red precipitate | Aldehyde (RCHO) Ketone (RCOR') | Aldehyde. I.R. $2720$ cm$^{-1}$, aldehyde $\nu$ C–H(w); $1725-1715$ cm$^{-1}$, R=Alk $\nu$ C=O (s, lowered by $\alpha\beta$-unsaturation to $\sim 1685$ cm$^{-1}$); $1700$ cm$^{-1}$, R = Ar $\nu$ C=O (s) N.M.R. $\delta$ 9.0–10.0; if *ortho* to –NO$_2$, –COOEt, –COOH or –COR, $\delta$ 9.0–12.0 Ketones. I.R. $1720-1710$ cm$^{-1}$, R and R' = Alk $\nu$ C=O (s, affected by $\alpha\beta$-unsaturation as for aldehydes); $1690$ cm$^{-1}$, R = Alk, R' = Ar $\nu$ C=O (s); $1665$ cm$^{-1}$, R and R' = Ar $\nu$ C=O (s) |
| a. Add 2 M ammonium hydroxide dropwise to 5% silver nitrate solution until the precipitate just dissolves. Add the organic compound and warm on a water bath. A little ethanol may be added if the compound is not water-soluble. | Silver mirror formed | Aliphatic aldehydes | N.B. Values for $\nu$ C=O in the I.R. spectra of aldehydes and ketones will be reduced if the group is concerned in hydrogen bonding, particularly with alcohol and phenol groups |

| Test procedure | Compound type | Observation | Spectra |
|---|---|---|---|
| b. If Test 2a is negative, treat 5% silver nitrate with 2 M sodium hydroxide (2 drops). Dissolve the precipitate obtained in the minimum of 2 M ammonium hydroxide. Add the organic compound and proceed as in Test 2a | Aromatic aldehyde | Silver mirror formed | |
| c. Tests 2a and 2b | Ketone | No silver mirror | |
| 3. (i) Add 5% chromium trioxide in 2 M sulphuric acid (3 drops). Warm at 40–50° for 1 min | Alcohol (AlkOH) | Red colour changes to green | I.R. $\sim 3300$ cm$^{-1}$, $\nu$ O–H (s, broad for most alcohols in solid or liquid state). If in dilute non-polar solution, $\nu$ O–H can be up to 3620 cm$^{-1}$ (no hydrogen bonding) N.M.R. $\delta$ 1.0–5.0 depending on solvent and concentration |
| (ii) Dissolve the compound in water, dioxan or propanone (acetone) and add 40% cerium(IV) nitrate in 2 M nitric acid (3 or 4 drops) | | Red coloration | |
| 4. If Test 3 is positive, and the unknown compound is a solid, dissolve in water and add 10% ethanolic 1-naphthol followed by slow addition (down side of test tube) of concentrated sulphuric acid | Carbohydrate | Violet coloration at liquid interface | I.R. Similar in many respects to the spectra of alcohols. Also 930–804 cm$^{-1}$, asym. ring vibration; 898–884 cm$^{-1}$, anomeric C–H equatorial deformation ($\beta$-sugars); 888–872 cm$^{-1}$, equatorial C–H deformation; 852–836 cm$^{-1}$, anomeric axial C–H deformation ($\alpha$-sugars) N.M.R. Complex signal $\delta$ 3.0–5.0 in D$_2$O |
| a. If Test 4 is positive, prepare a solution containing equal volumes of 0.1 M copper(II) sulphate and 0.1 M sodium potassium tartrate. Add 0.1 M sodium hydroxide until the precipitate just dissolves and then add the organic compound. Warm on a boiling-water bath for up to 5 min (Fehling's test). | Reducing carbohydrate | Brick-red precipitate | |
| b. Dissolve the compound in water and add a solution of 5% copper(II) ethanoate (acetate) in 1% aqueous ethanoic (acetic) acid. Boil gently for up to 2 min | Monosaccharide | Brick-red precipitate | |
| c. If Test 4b is positive, take the compound (0.1 g) and 1,3,5-trihydroxybenzene (phloroglucinol, 0.1 g) and dissolve in 2 M hydrochloric acid (2 cm$^3$). Heat to boiling for up to 2 min | Hexose Pentose | Pale yellow Intense red-brown coloration | |

**Table I.** Functional Groups Containing C, H, O (*cont.*)

| Chemical test | Observation | Functional group | Spectral data |
|---|---|---|---|
| 5. (i) Mix equal volumes of saturated methanolic hydroxylamine hydrochloride and saturated methanolic potassium hydroxide. Add the organic compound and heat on a boiling-water bath. Cool, acidify with 0.2 M hydrochloric acid and add 5% aqueous iron(III) chloride (3 drops). N.B. This test should not be applied if Tests 1b and 1c are positive | Red to violet coloration | Ester (RCO.OR') (and lactone) or Anhydride (RCO.O.COR) | Esters. I.R. 1735 cm$^{-1}$, R and R' = Alk $\nu$ C=O (s); 1725–1715 cm$^{-1}$, R = Ar or vinyl $\nu$ C=O (s); 1765–1755 cm$^{-1}$, R' = Ar or vinyl $\nu$ C=O (s)*; 1735 cm$^{-1}$, R and R' = Ar $\nu$ C=O (s); 1735 cm$^{-1}$, $\delta$-lactone $\nu$ C=O (s); 1770 cm$^{-1}$, $\gamma$-lactone $\nu$ C=O (s): 1300–1050 cm$^{-1}$, asym. and sym. $\nu$ C–O–C (2 bands, s) *N.B. Phthalate esters have $\nu$ C=O at 1780–1760 cm$^{-1}$ N.M.R. See Table XI |
| (ii) Dissolve the compound in ethanol; add Methanolic potassium hydroxide (1 drop) and phenolphthalein (1 drop). Prepare a similar mixture but omit the compound under examination. Heat both samples in a boiling-water bath | Pink colour fades in test solution | | Anhydrides. I.R. 1830–1810 cm$^{-1}$, 1770–1750 cm$^{-1}$, R = Alk $\nu$ C=O (doublet due to coupling, s); 1795–1775 cm$^{-1}$, 1735–1715$^{-1}$, R = Ar or vinyl $\nu$ C=O (doublet due to coupling, s); 1810–1790 cm$^{-1}$, 1760–1740 cm$^{-1}$, $\nu$ C=O (6-membered cyclic anhydride, s); 1875–1855 cm$^{-1}$, 1975–1775 cm$^{-1}$, $\nu$ C=O (5-membered cyclic anhydride) N.M.R. See comment relating to esters |
| a. If Test 5 is positive, dissolve the compound in benzene or trichloromethane (chloroform) and add aniline (2 drops). Warm gently for 1–2 min | Precipitate formed | Anhydride | |
| 6. (i) Visual examination | Compound has a red to yellow colour when pure | Quinone | I.R. 1690–1655 cm$^{-1}$, $\nu$ C=O N.M.R. $\delta$ 6.5–7.5 (protons in quinone ring) |
| (ii) Add 2 M sodium hydroxide | Pronounced intensification of the original colour | | |

| Test | Observation | Classification and spectral data |
|---|---|---|
| 7. (i) Warm at 40–50° with concentrated sulphuric acid<br><br>(ii) Apply Ferrox Test (p. 2) | Compound dissolves completely without charring<br><br>Violet coloration | Ether (ROR')<br>I.R. 1275–1200 cm$^{-1}$. R = Ar or vinyl asym. $\nu$ C–O–C (s); 1150–1070 cm$^{-1}$, R = Alk asym. $\nu$ C–O–C (s)<br>N.B. Absorptions in the regions given above are not necessarily conclusive evidence of an ether as such but only of the C–O–C grouping<br>N.M.R. See Table XI for effect of ether linkage on chemical shift of nearby protons |
| 8. (i) Treat with 1% potassium permanganate solution and shake well N.B. This test is not valid for compounds containing readily oxidized groups.<br>(ii) Dissolve the compound in carbon tetrachloride and add 5% bromine in carbon tetrachloride (2 drops) | Purple colour is discharged rapidly in cold<br><br>Red-brown colour is discharged *without evolution of hydrogen bromide* | Alkene (RCH=CHR')<br>Alkyne (RC≡CR')<br>Alkene I.R. 3100–3000 cm$^{-1}$, $\nu$ C–H (w–m); 1670–1610 cm$^{-1}$, $\nu$ C=C (variable intensity, absent if compound symmetrical about double bond); 1600–1590 cm$^{-1}$, $\nu$ C=C (conjugated, m); 1325–1275 cm$^{-1}$, $\gamma$ C–H(s); 900–880 cm$^{-1}$, rocking mode for C–H<br>N.M.R. $\delta$ 4.5–6.3<br>Alkyne I.R. 2260–2190 cm$^{-1}$, $\nu$ C≡C (disubstituted, w) 2140–2100 cm$^{-1}$, $\nu$ C≡C (monosubstituted, w).<br>N.M.R. $\delta$ 2.5–3.0 |
| 9. All of the above tests are negative | | Other hydrocarbons<br>I.R. 3080–3030 cm$^{-1}$, arom. $\nu$ C–H (often a triplet, w); 2960–2850 cm$^{-1}$, aliph. $\nu$ C–H(s); 1650–1450 cm$^{-1}$ arom. $\nu$ C=C best diagnostic peak at 1600 cm$^{-1}$, variable); 1465–1400 cm$^{-1}$, aliph. $\delta$ C–H (complex, m); 1275–960 cm$^{-1}$, arom. $\delta$ C–H; 1250–1200 cm$^{-1}$, 1170–1145 cm$^{-1}$, $\nu$ C–C (skeletal mode doublet, variable); < 900 cm$^{-1}$, arom. $\gamma$ C–H (important in interpretation of aromatic substitution pattern, see Table X)<br>N.M.R. Saturated alkanes $\delta$ 0.4–2.1; benzenoid $\delta$ 6.0–8.6 |

## Table II. Functional Groups Containing C, H, N, O

| Procedure | Observation | Functional group | Spectral data |
|---|---|---|---|
| 1a. Dissolve* in 2 M hydrochloric acid at room temperature, cool to 5° in ice and add 5% aqueous sodium nitrite (3 or 4 drops) to the acidic solution<br><br>*A few weakly basic amines require concentrated hydrochloric acid; if this fails, the amine should be dissolved in the minimum of ethanol and a little concentrated sulphuric acid should be added and the solution cooled in ice. Compounds which have a pyrrole ring are very weakly basic and do not give the above reactions | Effervescence; nitrogen is evolved and a clear solution is obtained | Pri.aliphatic amine (AlkNH₂), Amino acid ($^-$OOCAlkNH$_3^+$) or amide (RCONH₂) | Amines (Primary and Secondary, including hydrazines) I.R. 3500–3000 cm$^{-1}$, $\nu$ N–H (w–m, often a doublet for –NH₂ group due to asym. and sym. $\nu$ N–H); 1640–1560 cm$^{-1}$, $\delta$ N–H(s); 1360–1250 cm$^{-1}$, 1280–1180 cm$^{-1}$, $\nu$ C–N(doublet for unsaturated carbon); 1230–1030 cm$^{-1}$, $\nu$ C–N (v)*; 1150–1100 cm$^{-1}$, $\nu$ C–N (as in C–N–C, v); 900–650 cm$^{-1}$, $\gamma$ N–H (broad and diffuse)<br>N.B. Tertiary amines do not absorb in the region 3500–3000 cm$^{-1}$<br>*Doublet for tertiary amines<br>N.M.R. Aliphatic $\delta$ 1.5–3.5; benzenoid $\delta$ 3.5–5.5; heterocyclic $\delta$ 4.5–6.5. Saturated endocyclic NH $\delta$ 0.0–1.5. NH in pyrazole and in imidazole ring, $\delta$ 10.0–14.0. †Hydrazines show great variation in $\delta$ values<br>†NH of a pyrrole ring, $\delta$ 7.5–11.0<br>Amides I.R. 3300–3050 cm$^{-1}$, $\nu$ N–H (m, often a doublet); 1690–1650 cm$^{-1}$, $\nu$ C=O (Amide I band, s); 1640–1600 cm$^{-1}$, N–H bending mode (Amide II band); 1420–1405 cm$^{-1}$, $\nu$ C–N (Amide III band)<br>N.B. Most $N$-substituted amides show similar absorption to those given above for amides, but values for the Amides II and III bands are usually lowered by ~70–150 cm$^{-1}$; also $\nu$ N–H appears as a singlet.<br>Note that if R = Ar or vinyl, values are increased by ~15 cm$^{-1}$<br>N.M.R. $\delta$ 5.0–8.5, very broad for –CONHR, but usually sharp for –CONH₂. Two hydrogens of –CONH₂ are sometimes not equivalent and give two signals |
| | No effervescence; clear solution obtained | Pri-aromatic amine (ArNH₂) or tertiary amine (R₃N) | |
| | Dark brown soln. obtained | Tertiary aromatic amine unsubstituted in 4-position | |
| | No effervescence; cloudy solution or emulsion formed | Secondary amine (R₂NH) or arylhydrazine (Ar NHNH₂) | |
| b. If Test 1a gives a clear solution, add a few drops of this to a 5% solution of 2-naphthol dissolved in 2 M sodium hydroxide | No coloration; ignore white to yellow precipitates | Pri.aliphatic amine (AlkNH₂), tertiary amine(R₃N), amino acid ($^-$OOCAlkNH$_3^+$) or amide(RCONH₂) | |
| | Bright red to dark brown precipitate | Pri.aromatic amine (ArNH₂) | |
| c. Dissolve in 2 M hydrochloric acid at room temperature and add an excess of Fehling's solution. Heat on a boiling-water bath for 5 min | Brick-red precipitate | Substituted hydrazine(RNHNH₂). | |
| d. Dissolve in water and add 1% methanolic ninhydrin. Warm gently | Violet coloration | α- or β-Amino acid | |
| 2. Add concentrated sodium hydroxide solution. Heat strongly for about 2 min | Ammonia evolved | Ammonium salt (RCOONH$_4^+$), amide (RCONH₂), imide (RCONHCOHNR) or nitrile (RCN) | Amino acids. I.R. 3100–2600 cm$^{-1}$, $^+$NH₃ deformation (m); 1665–1585 cm$^{-1}$, $^+$NH₃ deformation (Amino acid I band, w); 1550–1485 cm$^{-1}$, $^+$NH₃ deformation (Amino acid II, $\nu$).<br>N.M.R. Usually insoluble in CDCl₃; spectra taken in D₂O which converts $^+$NH₃ to $^+$ND₃. |

Ammonium salts. I.R. 3300–3030 cm⁻¹, ν N–H (⁺NH₄ group); 1430–1390 cm⁻¹, bending mode for N–H (⁺NH₄ group) N.M.R. Usually insoluble in CDCl₃

Imides. I.R. Spectra are similar to those of N-substituted amides, particularly on low resolution instruments. Higher resolution can show a doublet for ν C=O.

N.M.R. δ 8.0–10.0, very broad.

N.B. The above –NH– or –NH₂ groups are converted into –ND– or ND₂ by treatment with D₂O

Nitriles. I.R. 2260–2240 cm⁻¹, R = Alk ν C≡N (m, very sharp); 2240–2220 cm⁻¹, R = Ar or vinyl ν C≡N.

Nitro compounds. I.R. 1615–1540 cm⁻¹, 1390–1320 cm⁻¹, R = Alk ν N=O (doublet for asym. and sym. modes, s); 1548–1508 cm⁻¹, R = Ar ν N=O (s, doublet as above).

Nitrophenols. I.R. and N.M.R. See individual functional group absorptions.

| Test | Observation | Conclusion |
|---|---|---|
| a. If Test 2 is positive, add cold 2 M sodium hydroxide solution. | Compound dissolves and ammonia is evolved | Ammonium salt |
| b. Mix with sulphur in a dry test tube and heat gently. Test the vapour produced with an absorbent strip treated with 1% aqueous iron(III) nitrate solution | Red stain on paper | Nitrile |
| c. If Test 2 is positive but Tests 2a and 2b are negative, an imide may be distinguished from an amide by mixing saturated methanolic solutions of the compound and potassium hydroxide | White precipitate formed | Imide |
| 3. Add 70% sulphuric acid and reflux for 10 min. Cool in ice to 5° and add 5% sodium nitrite solution. Pour this mixture into a 5% solution of 2-naphthol in 2 M sodium hydroxide. | Bright red to dark brown precipitate | N-Substituted amide (RCONHR) |
| 4. (i) To an aqueous solution of iron(II) sulphate solution (1 cm³) add a few drops of 2 M sodium hydroxide followed by the organic compound. Prepare a similar mixture but omit the organic compound. Heat on a boiling-water bath for not more than 2 min. (ii) Dissolve the compound in propanone(acetone) and add 5% titanium(III) chloride or sulphate solution (3–5 drops); warm gently | Grey to green precipitate in test solution changes to brown / Mauve color is discharged within 2 min | Nitro compound (RNO₂) |
| a. If Test 4 is positive, reduce the compound as follows: add tin and 7 M hydrochloric acid and warm, with continual shaking, for 15 min. Filter the mixture, cool to 5° in ice and add 5% aqueous sodium nitrite. Add this to a 5% solution of 2-naphthol in 2 M sodium hydroxide. | Red to dark brown precipitate | Aromatic nitro compound (ArNO₂) |
| | No coloured precipitate; ignore white to yellow precipitate | Aliphatic nitro compound (AlkNO₂) |
| b. Add 2 M sodium hydroxide solution | Intense yellow or orange coloration or precipitate | Nitrophenol (HOArNO₂) |

**Table III.** Functional Groups Containing C, H, O, Halogen

| | | | |
|---|---|---|---|
| 1. Prepare a clear saturated solution of the compound in 2 M nitric acid and add aqueous silver nitrate | White precipitate | Acyl chloride(RCOCl) | I.R. 1810–1780 cm$^{-1}$, $\nu$ C=O (doublet when R = Ar, s) |
| 2. If Test 1 is negative, make a solution of the compound in ethanol and add 2% ethanolic silver nitrate | White to yellow precipitate formed in cold or on slight warming | Alkyl halide (AlkX, where X = Cl, Br, I, F) | I.R. 1250–960 cm$^{-1}$, $\nu$ C–F; 830–500 cm$^{-1}$, $\nu$ C–Cl (s, note overtone band at 1510–1480 cm$^{-1}$) N.B. Stretching frequencies for C–Br and C–I are normally outside the range of simple infrared spectrophotometers |
| Tests 1 and 2 are negative | | Aryl halide (ArX, where X = Cl, Br, I, F) | |

**Table IV.** Functional Groups Containing H, O, S

| | | | |
|---|---|---|---|
| 1. Add water, shake well and test with blue litmus paper | Readily soluble with acid reaction | Sulphonic acid (RSO$_2$OH) | Sulphonic acids. I.R. 1250–1160 cm$^{-1}$, 1080–1000 cm$^{-1}$, $\nu$ S=O (doublet due to asym. and sym. stretching modes); 700–610 cm$^{-1}$, $\nu$ S–O. Note that absorption of the O–H group is similar to that in carboxylic acids |
| 2. Odour | Unpleasant and penetrating | Thiol or thiophenol (RSH) (also impure thioether) | N.M.R. Usually insoluble in CDCl$_3$; in water, $\delta$ 11.0–12.0 Thiols and Thiophenols. I.R. 2600–2500 cm$^{-1}$, $\nu$ S–H (s) |
| a. If Test 2 is positive, dissolve in ethanol and add solid sodium nitrite followed by 2 M sulphuric acid | Red coloration | Pri.or secondary thiol (AlkSH) | N.M.R. Similar to alcohols and phenols (Table I). |
| | Green coloration changing to red on standing | Thiophenol (ArSH) | Thioethers. I.R. 695–655 cm$^{-1}$, 630–600 cm$^{-1}$, C–S–C (doublet) N.M.R. Similar to ethers |
| | No coloration | Thioether (RSR) | |

16

**Table V.** Functional Groups Containing O, P

| | | | |
|---|---|---|---|
| 1. Reflux with 30% aqueous sodium hydroxide for 20 min and then distil off all the volatile material. Acidify the residue with 2 M sulphuric acid, extract with ether and add a solution of ammonium molybdate in concentrated nitric acid to the aqueous phase. Warm but do not boil | Yellow precipitate obtained | Phosphate ester $(R_3PO_4)$ | I.R. 1315–1180 cm$^{-1}$, $\nu$ P=O (s); 1195–1185 cm$^{-1}$, $\nu$ P=O (w, sharp if R = CH$_3$); 1100–950 cm$^{-1}$, $\nu$ P–O–C; 950–875 cm$^{-1}$, $\nu$ P=O (w, (s if R = Ph) |

**Table VI.** Functional Groups Containing H, Halogen, N

| | | | |
|---|---|---|---|
| 1. Dissolve in water; add excess of nitric acid and then aqueous silver nitrate | White to yellow precipitate formed | Hydrohalide salt of a base $(R\overset{+}{N}H_3\ \bar{X},\ R_2\overset{+}{N}H_2\bar{X},\ R_3\ \overset{+}{N}H\bar{X})$ or quaternary ammonium salt $(R_4\overset{+}{N}\bar{X})$ | Hydrohalides. I.R. 3000–2250 cm$^{-1}$, $\nu$ N–H (often very broad and complex, particularly for salts of secondary and tertiary amines); 1600–1575 cm$^{-1}$, N–H bending mode (doublet with band at 1500 cm$^{-1}$ for salts of primary amines). Note that these bands are absent for salts of tertiary amines N.M.R. Usually insoluble in CDCl$_3$, and should be converted into the free amines (q.v.) |
| a. If Test 1 is positive, add excess of alkali and extract the mixture with ether. Dry the extract over anhydrous sodium sulphate and evaporate the ether on a water bath | Residue obtained; examine as described in Table II. Tests 1a, 1b and 1c | Hydrohalide salt of a base | Quaternary ammonium salts. I.R. No characteristic absorptions N.M.R. Usually insoluble in CDCl$_3$ |
| | No residue | Quaternary ammonium salt | |

**Table VII.** Functional Group Containing Halogen, O, S

| | | | |
|---|---|---|---|
| 1. Heat with water for 10 min and cool. Acidify with 2 M nitric acid and add aqueous silver nitrate | White to yellow precipitate | Sulphonyl halide ($RSO_2Cl$) | Sulphonyl Halides. I.R. $1385-1320$ cm$^{-1}$, $1185-1150$ cm$^{-1}$, $\nu$ S=O (doublet for asym. and sym. modes) |

**Table VIII.** Functional Groups Containing C, H, N, O, S

| | | | |
|---|---|---|---|
| 1. Heat with solid potassium hydroxide | Ammonia evolved | Sulphonamide ($RSO_2NH_2$) | I.R. $1375-1330$ cm$^{-1}$, $1180-1160$ cm$^{-1}$, S=O (doublet for asym. and sym. modes). Other absorptions are similar to those of amides and N-substituted amides |
| | Amine evolved | N-Substituted sulphonamide ($RSO_2NHR'$ or $RSO_2NR^1R^2$) Examine as described in Table II Tests 1a and 1b | |
| 2. Heat with 2 M sulphuric acid | Hydrogen sulphide evolved | Thioamide ($RCSNH_2$) | I.R. $1405-1290$ cm$^{-1}$, C=S (Analogous to Amide 1 band for amides, see Table II) |
| 3. Dissolve in water, acidify with 2 M hydrochloric acid and add barium chloride solution | White precipitate | Sulphate salt of base ($R\overset{+}{N}H_3\ \bar{H}SO_4, R_2\overset{+}{N}H_2\ \bar{H}SO_4$ or $R_3\overset{+}{N}H\ \bar{H}SO_4$) | I.R. See data for hydrohalides of bases (Table VI) |

**Table IX.** Functional Group Containing H, N, O. P

| | | | |
|---|---|---|---|
| 1. Dissolve in water and treat with concentrated nitric acid and aqueous ammonium molybdate. Warm, but do not boil | Yellow precipitate obtained | Phosphate of a base | I.R. See data for hydrohalides of bases (Table VI) |

**Table X.** Out-of-plane bending modes ($\gamma$ C—H) values for substituted benzene ring

| Substitution | Frequency (cm$^{-1}$) | Intensity |
|---|---|---|
| Monosubstitution | 900—860 | w—m |
| | 770—730 | s |
| | 710—690 | s |
| 1,2-Disubstitution | 960—905 | w |
| | 850—810 | w |
| | 760—745 | w |
| 1,3-Disubstitution | 960—900 | m |
| | 880—830 | m—s |
| | 820—790 | w—m |
| 1,4- and 1,2,3,4- Substitution | 860—800 | s |
| 1,2,3-Trisubstitution | 965—950 | w |
| | 900—885 | w |
| | 780—760 | s |
| | 720—686 | m |
| 1,2,4-Trisubstitution | 940—920 | w |
| | 900—885 | m |
| | 780—760 | s |
| 1,3,5-Trisubstitution | 950—925 | v |
| | 860—830 | s |
| 1,2,3,5-, 1,2,4,5- and 1,2,3,4,5-Substitution | 970—850 | s |

**Table XI.** Chemical shift of protons attached to carbon

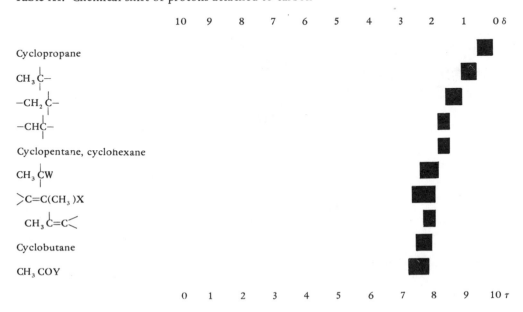

**Table XI.** Chemical shift of protons attached to carbon (*cont.*)

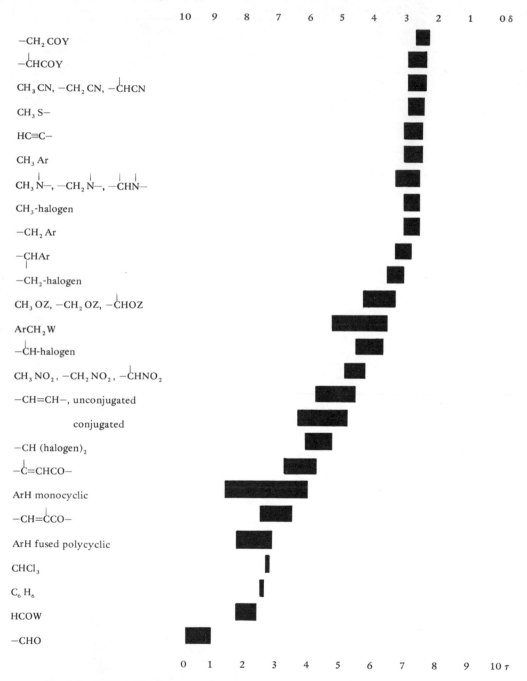

W = halogen, OH, OR, OAr, $NH_2$, $NR_2$, NHR
X = CHO, COR, $CO_2R$, $CO_2H$, OCOR
Y = H, halogen, OH, OR, $NH_2$, NHR, $NR_2$
Z = H, R, Ar, COR

**Table XII.** Chemical shift of protons attached to oxygen, nitrogen or sulphur

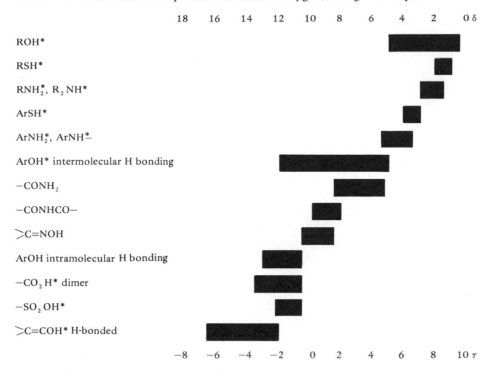

*Chemical shift is dependent on concentration and temperature

21

# 3

## THE SEPARATION OF ORGANIC MIXTURES

A mixture of organic compounds may be in the solid or liquid form or may consist of a solid dissolved or suspended in a liquid. If a solid and a liquid are present it is usually unwise to expect separation to be accomplished by filtration because the liquid phase almost certainly contains some dissolved solid and traces of the liquid component may be difficult to remove from the solid compound. The methods of isolating pure samples of the components from a mixture may be either physical or chemical. The physical method consists of fractional distillation and is applicable only if there is a wide difference between the boiling points of the two compounds and provided that an azeotrope is not formed. The chemical method of separating two compounds depends on their differing solubility in water, ether, dilute acid or alkali.

The procedures described below should be followed in the order given and should be successful for the great majority of mixtures.

1. If the mixture is a liquid, it should be placed in a small flask equipped with a stillhead, thermometer and condenser. Heat the flask carefully; observe whether a liquid distils, and if it does, note the temperature and continue the distillation until the temperature at the stillhead falls, indicating that all the liquid boiling at that temperature has distilled over. The lower-boiling component of the mixture will now be in the receiver flask.

2. If the mixture is a liquid which cannot be separated according to paragraph 1, or if the mixture is a solid, test its solubility in ether. Most organic compounds are soluble but amongst those which have a low solubility are the following: carbohydrates, amino-acids, sulphonic acids, salts of amines, metal salts of carboxylic and sulphonic acids, some aromatic polybasic acids, some amides and ureas, and polyhydroxy compounds.

*Liquid mixtures:* proceed to test (b) below.

*Solid mixtures:* if the mixture dissolves completely, proceed to test (b) below. If an insoluble part remains, continue as described in (a). If both components are insoluble, use the sequence of tests given in paragraph 5 below.

(a) Filter off the undissolved solid, keep the ethereal filtrate (A). Allow the solid collected to dry off in air or over slight heat.

Take the filtrate (A) and if all the mixture did not dissolve in ether initially, evaporate the ether over a hot water bath. If a residue (solid or liquid) is obtained, a separation has been achieved by virtue of the ether-solubility of one of the two compounds.

(b) If all the mixture dissolved in ether, place the solution in a tap funnel. Extract this with 10% sodium hydroxide solution, separate the basic extract from the ether layer (B) into a small flask and acidify with 10% hydrochloric acid solution. The appearance of a solid or an oil or of turbidity may indicate the presence of a carboxylic acid or a phenolic compound.

Take the ether layer B and extract it with 10% hydrochloric acid solution. Separate the acid layer (keep the ethereal solution (C)) and basify it with dilute sodium hydroxide. If an oil or a solid separates, extract this twice with ether. Dry the extract with anhydrous sodium sulphate, allow to stand for 10 min in a stoppered flask, evaporate the ether over a hot-water bath and the residue, if any, will be the basic component.

The ethereal layer (C) contains a neutral compound if present in the original mixture; it should be dried with sodium sulphate, filtered and the solvent distilled. The most common classes of neutral compounds are: hydrocarbons, ethers, halides, alcohols, aldehydes, ketones, amides, nitriles, esters, unreactive anhydrides and nitro compounds.

If basic and neutral components are absent, a carboxylic acid and a phenol may be separated as follows: to this mixture add an excess of solid sodium bicarbonate in small amounts with constant stirring until the solution is no longer acid to litmus. Extract the solution with ether; the ethereal layer will contain a phenolic compound if present while the aqueous layer will contain a carboxylic acid.* Acidify this aqueous solution with dilute hydrochloric acid. A solid carboxylic acid may separate out and should be filtered off and dried. If a liquid acid is present, some solid calcium chloride should be added with thorough shaking until the water is almost saturated with it. The acid is then extracted with ether and isolated in the usual way.

3. The procedure given in paragraph 2 above will not separate two neutral compounds. If nothing is extracted from the ethereal solution by dilute acid and dilute alkali, the presence of two neutral compounds should be suspected. If one of these is a carbonyl compound which forms a bisulphite adduct, this may be removed as its addition product. Prepare a 40% solution of sodium metabisulphite in water and add one fifth of its volume of ethanol. If some of the salt separates, filter the solution and to the filtrate (12 cm$^3$) add the mixture (4 g). Shake thoroughly and allow to cool. The crystalline adduct if formed should be filtered off (keep the filtrate (D)) and decomposed by careful distillation with an excess of sodium carbonate solution. The carbonyl compound if volatile is found in the distillate; otherwise it will have to be extracted out of the mixture in the distilling flask with ether and isolated in the usual way.

The other neutral component will be found in the filtrate D and should be isolated by removing any ethanol that may remain and then extracted with ether.

4. A few compounds which are very soluble in water and in ether can be troublesome to isolate from mixture. For example, 1,3-dihydroxybenzene mixed with an amine or a carbonyl compound would not be readily separated by the procedure given in paragraph 2. 1,3-Dihydroxybenzene is about equally soluble in water as in ether and although it would be extracted out of ether with several portions of dilute sodium hydroxide solution, acidification of this extract will not result in separation of the compound because of its solubility in water. A mixture consisting of an amine and resorcinol may be separated as follows.

Dissolve the mixture in an excess of dilute hydrochloric acid; test the solution with litmus paper. Place the solution in a tap funnel and extract it with three portions of ether each equal to about one half the volume of the mixture. Retain the acidic solution (E). Dry the ethereal extract with anhydrous sodium sulphate, filter and distil off the solvent. The residue should be tested for a phenolic group. Basify the acidic solution E with sodium hydroxide solution, extract with ether. Dry the extract and distil off the solvent. A residue indicates the presence of an amine.

A water-soluble phenol or polyhydric phenol and a carbonyl compound are best separated by converting the latter into its bisulphite compound (see paragraph 3) or its semicarbazone

---

*Some nitro- and halogeno-phenols are sufficiently acidic to release carbon dioxide from sodium bicarbonate. Therefore an aromatic compound which contains nitrogen or a halogen and which behaves as carboxylic acid in this separation should be tested for phenolic properties.

and regenerating it by heating with twice its weight of oxalic acid and ten times its weight of water in a distilling flask. The carbonyl compound, if low boiling, may be distilled and collected, or if high boiling, may be extracted with ether from the residue in the distilling flask and characterized in the usual way.

5. The procedures given above will not be effective if both components are insoluble in ether. In such cases the following suggestions should be useful:

(a) Test the mixture to see if one component is soluble in water. If this is so, filter off the insoluble compound and evaporate the filtrate to dryness on a small Bunsen flame. Overheating may cause decomposition or charring. If a carbohydrate is present a syrup will often be obtained which may be difficult to crystallize but tests should be made on the syrup to determine its nature.

(b) Should both components be insoluble in water or both soluble in water, repeat procedure 5(a) using methanol.

(c) If the above tests have failed to effect a separation, the presence of any two of the following should be suspected: polyhydric alcohol, carbohydrate, metallic salt of a carboxylic acid or a salt of an organic base. Dissolve the mixture in 10% hydrochloric acid and if a solid precipitates, collect on a filter, wash with water and dry carefully. This may be a free acid, probably aromatic. Formation of an oil will indicate a higher aliphatic acid. The aqueous layer may contain a soluble polyol or carbohydrate.

If no solid or oil is obtained, extract the solution several times with ether and evaporate the extract on a water bath. A residue may be a lower aliphatic acid. If nothing was extracted by the ether, make the solution basic with 10% sodium hydroxide; a salt of an organic base if present will lead to the formation of the free base which should be extracted with ether and isolated in the usual way. The aqueous layer may contain a polyol or a carbohydrate together with sodium chloride or sulphate. Removal of the inorganic ions by passing the solution through a mixed-bed ion-exchange column should give an aqueous solution of the polyol or carbohydrate.

# 4

## PREPARATION OF DERIVATIVES

The identity of an organic compound may be confirmed by converting it into a crystalline derivative of characteristic melting point. The derivative should preferably have a melting point not lower than about 100° because such compounds are usually more readily crystallized and dried.

### Purification of derivatives

When preparing a derivative, it is necessary that the product be obtained in a pure form, thus ensuring an accurate melting point. The procedure for recrystallization is as follows: the crude material is dissolved in the minimum of hot solvent, filtered if necessary without suction, and the solution obtained allowed to cool slowly. The crystalline product is collected in a Buchner funnel, washed with a small volume of ice-cold solvent, dried thoroughly and the melting point determined. This procedure should be repeated until no further increase in melting point is observed.

If the crude material is badly discoloured, it is often advantageous to add a little activated charcoal to the solution which should then be boiled gently for up to 5 min. The hot solution is filtered without suction and the filtrate allowed to cool slowly.

The choice of a suitable solvent for recrystallization is of considerable importance in the purification of derivatives. Ideally, the compound should have a high solubility in the hot solvent and a low solubility in the cold solvent. If two or more solvents meet this requirement, that with the lowest boiling point should be chosen to facilitate removal from the solid product. It is sometimes impossible to find a satisfactory solvent and then the technique of mixed solvents is often of value. The derivative is dissolved in a slight excess of hot solvent and a second solvent which is miscible with the first solvent but in which the derivative is insoluble, is carefully added with shaking and warming until a faint cloudiness persists. This should disappear on boiling and the solution is allowed to cool slowly.

Experimental details are given below for preparing the derivatives listed in Chapter 5 for the various classes of organic compounds.

## ACETALS

### (a) Hydrolysis to aldehyde and alcohol

$$RCH(OAlk)_2 \xrightarrow{\text{aq. acid}} RCHO + 2\ AlkOH$$

$$AlkOH + CS_2 + NaOH \longrightarrow AlkOCS_2^- Na^+$$

Treat the acetal (0.5 g) with M hydrochloric acid (5 cm$^3$) and reflux for up to 30 min. If the resulting solution is homogeneous, divide into two equal portions and characterize the aldehyde as the 2,4-dinitrophenylhydrazone (see Aldehydes, p. 27) and the alcohol as the potassium

25

alkyl xanthate as follows. To a portion from the hydrolysis, add solid potassium hydroxide (7 g) and cool to 40°. Transfer to a separating funnel and add carbon disulphide (3 cm³) (INFLAMMABLE) and propanone (acetone) (3 cm³). Mix cautiously and then shake vigorously for 15 min. Allow to settle, discard the lower layer and filter the remaining solution through glass wool. Precipitate the xanthate with ether and recrystallize from ethanol.

If two layers separate, they should be treated individually as described above.

## ALCOHOLS

### (a) Ethanoate (acetate)

$$AlkOH + (CH_3 \cdot CO)_2O \longrightarrow AlkO \cdot CO \cdot CH_3 + CH_3 \cdot CO_2H$$

To the alcohol (0.5 g) add anhydrous sodium ethanoate (acetate) (0.5 g) and ethanoic (acetic) anhydride (3 cm³). Reflux for 20 min and pour into water (25 cm³). Stir until a solid is obtained, filter this off and wash well with water. Recrystallize from ethanol.

### (b) Benzoate and toluene-4-sulphonate (Schotten–Baumann method)

$$AlkOH + C_6H_5 \cdot COCl \longrightarrow AlkO \cdot CO \cdot C_6H_5 + HCl$$

$$AlkOH + 4-CH_3 \cdot C_6H_4 \cdot SO_2Cl \longrightarrow AlkO \cdot SO_2 \cdot C_6H_4 \cdot 4-CH_3 + HCl$$

Dissolve the alcohol (0.5 g) in 2 M sodium hydroxide (10 cm³) and add benzoyl chloride (1 cm³) or toluene-4-sulphonyl chloride (1 g). In the latter case, add just sufficient propanone (acetone) to render the mixture homogeneous. Shake vigorously in a stoppered tube until a solid is obtained. If propanone has been used, it may be necessary to add water in order to precipitate the product. Filter this off, wash well with water and recrystallize from ethanol.

### (c) Benzoate and toluene-4-sulphonate (alternative procedure); 4-nitrobenzoate and 3,5-dinitrobenzoate

$$AlkOH + ArCOCl \longrightarrow AlkO \cdot COAr + HCl$$

where Ar = $C_6H_5-$, $4-CH_3 \cdot C_6H_4SO_2-$, $4-NO_2 \cdot C_6H_4-$ or $3,5-(NO_2)_2 \cdot C_6H_3-$

Dissolve the alcohol (0.5 g) in dry pyridine (5 cm³) and add 4-nitrobenzoyl chloride (1 g) or 3,5-dinitrobenzoyl chloride (1.3 g). Reflux for 30 min and pour into 2 M hydrochloric acid (40 cm³). Separate the solid (sometimes an oil is formed) and stir with M sodium carbonate solution (10 cm³). Filter off the solid obtained and recrystallize from ethanol, aqueous ethanol or benzene.

### (d) Alkyl hydrogen 3-nitrophthalate

Heat a mixture of the alcohol (0·5 g) and 3-nitrophthalic anhydride (0·5 g) until a liquid is obtained. Continue heating for a further 15 min, cool, and recrystallize the solid obtained from water or aqueous ethanol. If the original alcohol has a boiling point in excess of $150°$, it is advisable to dissolve the mixture in toluene ($2-5$ cm$^3$) and reflux for up to 30 min. If no solid product is obtained on cooling, precipitate the product by addition of light petroleum ($60-80°$).

### (e) Oxidation of primary alcohols* to carboxylic acid

$$RCH_2 \cdot OH \xrightarrow{[O]} RCO_2H$$

Treat the alcohol (1 g) with a chromic acid oxidation mixture (10 cm$^3$) consisting of 50% sodium dichromate solution in 6 M sulphuric acid. Reflux until the colour has changed from red to green and add more oxidizing mixture. Continue this procedure until no further discharge of the red colour is observed. Cool the solution, filter off the resulting product and wash well with M sulphuric acid and then with water. Dissolve the product in sodium carbonate solution, filter, and acidify with M sulphuric acid. Filter off the solid, wash well with water and recrystallize from water or ethanol.

# ALDEHYDES

### (a) 2,4-Dinitrophenylhydrazone

$$RCHO + 2,4\text{-}(NO_2)_2C_6H_3 \cdot NH \cdot NII_2 \longrightarrow RCH:N \cdot NH \cdot C_6H_3(NO_2)_2 + H_2O$$

To the compound (0.5 g) dissolved in ethanol (0.5 cm$^3$) add a solution (2 cm$^3$) of 2,4-dinitrophenylhydrazine (see below) and boil for 2 min. Filter off the resulting precipitate, wash with a little cold ethanol and recrystallize from ethanol, ethanoic (acetic) acid, ethyl ethanoate (acetate) or trichloromethane (chloroform).

### Preparation of reagent

(i) Prepare a saturated solution in either 5 M hydrochloric acid or 3 M sulphuric acid.
(ii) Dissolve the 2,4-dinitrophenylhydrazine (2 g) in methanol (30 cm$^3$) and water (10 cm$^3$). Add concentrated sulphuric acid (4 cm$^3$) cautiously with shaking. Cool and filter if necessary.
(iii) Dissolve 2,4-dinitrophenylhydrazine in a solution of 85% phosphoric acid (60 cm$^3$) in ethanol (40 cm$^3$), heating gently if necessary.

### (b) Semicarbazone

$$RCHO + H_2N \cdot NH \cdot CO \cdot NH_2 \longrightarrow RCH:N \cdot NH \cdot CO \cdot NH_2 + H_2O$$

To a solution of the compound (0.5 g) in water (2 cm$^3$) add hydrated sodium ethanoate (acetate) (0.75 g) and semicarbazide hydrochloride (0.5 g). Add ethanol dropwise if the solution is not completely clear but care should be taken to add the minimum amount of ethanol otherwise sodium chloride may be precipitated. Heat for up to 10 min on a boiling-water bath, cool and filter. Recrystallize the product from ethanol, water, benzene or ethanoic (acetic) acid.

*This method may also be used to oxidize alkylbenzenes: $ArAlk \xrightarrow{[O]} ArCO_2H$

### (c) Oxime

$$RCHO + H_2N \cdot OH \longrightarrow RCH:N \cdot OH + H_2O$$

Dissolve hydroxylamine hydrochloride (0.5 g) in water (3 cm³) and add hydrated sodium ethanoate (acetate) (0.5 g) followed by the aldehyde (0.5 g). Heat on a boiling-water bath and add ethanol dropwise, if necessary, to clear the solution. Continue heating for 1–2 hours and allow to cool. Filter off the solid and recrystallize from ethanol.

### (d) 4-Nitrophenylhydrazone (and phenylhydrazone)

$$RCHO + 4\text{-}NO_2 \cdot C_6H_4 \cdot NH \cdot NH_2 \longrightarrow RCH:N \cdot NH \cdot C_6H_4 \cdot 4\text{-}NO_2 + H_2O$$

Prepare a solution of 4-nitrophenylhydrazine (0.5 g) in ethanol (15 cm³) and ethanoic (acetic) acid (0.5 cm³) and add the organic compound (0.5 g) to the solution. Reflux for 10 min, cool and recrystallize the solid product from ethanol. Occasionally, no solid is obtained on cooling. In such cases, the solution should be reheated and water added until a faint cloudiness is observed. Recrystallize the product using this technique (mixed solvents).

For the preparation of phenylhydrazones, replace the 4-nitrophenylhydrazine with phenylhydrazine (0.5 g).

### (e) Dimethone

Treat the aldehyde (0.5 g) with a 5% solution of 1,1-dimethylcyclohexane-3,5-dione (dimedone) in 50:50 aqueous ethanol (10 cm³). If no precipitate forms within 2 min, warm the solution for 5 min, cool in ice and filter off the product. Recrystallize from aqueous ethanol or ethanol.

## AMIDES, IMIDES, UREAS AND GUANIDINES

### (a) Xanthyl derivative

To the organic compound (0.5 g) add a 7% solution of xanthydrol in ethanoic (acetic) acid (7 cm³) and reflux for up to 30 min. Add water (5 cm³) and allow to cool. Filter off the solid product and recrystallize from aqueous dioxan or ethanoic acid.

### (b) Hydrolysis to carboxylic acid

$$RCO \cdot NH_2 + NaOH \longrightarrow RCO_2Na + NH_3 \xrightarrow{\text{aq. acid}} RCO_2H$$

28

Reflux the organic compound (0.5 g) with an excess of 6 M sodium hydroxide solution until no further evolution of ammonia is detectable. Acidify the resulting solution with concentrated hydrochloric acid and filter off the solid product obtained. Wash well with water and recrystallize from water, aqueous ethanol or ethanol.

## AMIDES, N-SUBSTITUTED

### (a) Hydrolysis to carboxylic acid and amine

$$RCO \cdot NHR' \xrightarrow{\text{aq. acid}} RCO_2H + R'NH_2$$

Treat the organic compound (1 g) with 12 M sulphuric acid (4 cm$^3$) and reflux for 30 min. Cool, basify with sodium hydroxide solution and extract the liberated base with ether. Obtain the free base from the ethereal extract by evaporating the ether on a boiling-water bath and characterize as described under *Amines, Primary and Secondary* (see below). Acidify the remaining aqueous solution with hydrochloric acid and filter off any solid produced. Recrystallize from water, aqueous ethanol or ethanol to obtain the pure carboxylic acid and characterize as described under *Carboxylic Acids* (p. 32). If no solid separates, saturate the solution with sodium chloride and extract with ether. Evaporate the ether and characterize the residue as described under *Carboxylic Acids* (p. 32).

## AMINES, PRIMARY AND SECONDARY

### (a) Ethanoyl (acetyl) derivative

$$RNH_2 + (CH_3 \cdot CO)_2O \longrightarrow RNH \cdot CO \cdot CH_3 + CH_3 \cdot CO_2H$$

$$RR'NH + (CH_3 \cdot CO)_2O \longrightarrow RR'N \cdot CO \cdot CH_3 + CH_3 \cdot CO_2H$$

Suspend the amine (0.5 g) in water (0.5 cm$^3$) and add a mixture of ethanoic (acetic) acid (0.5 cm$^3$) and ethanoic (acetic) anhydride (0.5 cm$^3$). Heat gently if reaction does not occur immediately. Cool and filter off any solid which separates. If no solid is obtained, neutralize the solution with saturated sodium carbonate solution. Filter off the solid obtained. In either case, recrystallize the product from water or aqueous ethanol. To ethanoylate (acetylate) weakly basic amines, replace the water by a few drops of concentrated sulphuric acid and reflux for 20 min.

### (b) Benzoyl, benzenesulphonyl and toluene-4-sulphonyl derivatives

$$RNH_2 + C_6H_5 \cdot COCl \longrightarrow RNH \cdot CO \cdot C_6H_5 + HCl$$

$$RR'NH + C_6H_5 \cdot COCl \longrightarrow RR'N \cdot CO \cdot C_6H_5 + HCl$$

$$RNH_2 + ArSO_2Cl \longrightarrow RNH \cdot SO_2Ar + HCl$$

$$RR'NH + ArSO_2Cl \longrightarrow RR'N \cdot SO_2Ar + HCl$$

where Ar = $C_6H_5-$ or $4-CH_3C_6H_4-$.

Prepare as described under *Alcohols* (p. 26).

**(c) 2,4-Dinitrophenyl derivative**

$$RNH_2 + 2,4\text{-}(NO_2)_2C_6H_3Cl \longrightarrow 2,4\text{-}(NO_2)_2C_6H_3 \cdot NHR + HCl$$

$$RR'NH + 2,4\text{-}(NO_2)_2C_6H_3Cl \longrightarrow 2,4\text{-}(NO_2)_2C_6H_3 \cdot NRR' + HCl$$

Treat the amine (0.5 g) with an equimolar proportion of 2,4-dinitrochlorobenzene (CAUTION: skin irritant) and anhydrous sodium ethanoate (acetate) (1 g); heat on a boiling-water bath for up to 30 min. Cool, and add cold ethanol (3—4 cm$^3$). Filter off the solid obtained and recrystallize from ethanol.

**(d) Picrate**

$$RR'R''N + 2,4,6\text{-}(NO_2)_3C_6H_2 \cdot OH \longrightarrow [RR'R''NH]^+[2,4,6\text{-}(NO_2)_3C_6H_2O]^-$$

where R$'$ and/or R$''$ may be a hydrogen atom.

Dissolve the amine (0.5 g) in ethanol (2 cm$^3$) and treat with a saturated ethanolic solution of picric acid (3 cm$^3$). Warm gently for 1 min and allow to cool. Recrystallize the product from ethanol.

# AMINES, TERTIARY

**(a) Methiodide**

$$RR'R''N + CH_3I \longrightarrow [RR'R''N \cdot CH_3]^+[I]^-$$

Add methyl iodide (0.5 cm$^3$) to the dry amine (0.5 g) at room temperature and allow to stand for 5 min. Reflux on a boiling-water bath for a further 5 min and then cool in ice. Filter off the solid product (scratch with a glass rod if no solid is obtained) and recrystallize from ethanol or propanone (acetone).

**(b) Picrate**

$$RR'R''N + 2,4,6\text{-}(NO_2)_3C_6H_2 \cdot OH \longrightarrow [RR'R''NH]^+[2,4,6\text{-}(NO_2)_3C_6H_2O]^-$$

Prepare as described under *Amines, Primary and Secondary* (above).

**(c) 4-Nitroso derivative** (for dialkyl tertiary amines with vacant 4-position in aryl group)

$$R_2N \cdot C_6H_5 + HONO \longrightarrow 4\text{-}R_2N \cdot C_6H_4 \cdot NO + H_2O$$

Dissolve the amine (0.5 cm$^3$) in 2 M hydrochloric acid (4 cm$^3$) and cool in ice to 5°. Add dropwise 20% sodium nitrite solution (2 cm$^3$) and allow to stand in the cold for 5 min. Basify with 2 M sodium hydroxide and extract with trichloromethane (chloroform). Dry the extract over anhydrous sodium sulphate and precipitate the derivative by addition of carbon tetrachloride. Filter off the product and recrystallize from ether (INFLAMMABLE).

**(d) Methyl toluene-4-sulphonate salt**

$$RR'R''N + 4\text{-}CH_3 \cdot C_6H_4 \cdot SO_2 \cdot O \cdot CH_3 \longrightarrow [RR'R''N \cdot CH_3]^+[4\text{-}CH_3 \cdot C_6H_4 \cdot SO_2 \cdot O]^-$$

To the amine (0.2 cm$^3$) add methyl toluene-4-sulphonate (0.3 g) and benzene (1 cm$^3$) or diethyl ether (1 cm$^3$). Reflux on a water bath for 20 min and cool. Decant the ether (crystals should remain) and add methanol (1 cm$^3$) and ethyl ethanoate (acetate) (5 cm$^3$) for recrystallization.

## AMINO-ACIDS

**(a) Benzoyl, 3,5-dinitrobenzoyl and toluene-4-sulphonyl derivatives**

$$\begin{array}{c}CO_2H\\|\\R\cdot NH_2\end{array} + ArCOCl \longrightarrow \begin{array}{c}CO_2H\\|\\R\cdot NH\cdot COAr\end{array} + HCl$$

where Ar = $C_6H_5-$ or $3,5\text{-}(NO_2)_2C_6H_3-$.

$$\begin{array}{c}CO_2H\\|\\R\cdot NH_2\end{array} + 4\text{-}CH_3\cdot C_6H_4\cdot SO_2Cl \longrightarrow \begin{array}{c}CO_2H\\|\\R\cdot NH\cdot SO_2\cdot C_6H_4\cdot 4\text{-}CH_3\end{array}$$

Prepare as described under *Amines, Primary and Secondary* (using 3,5-dinitrobenzoyl chloride (1 g) to prepare the 3,5-dinitrobenzoyl derivative). In each case acidify the solution with 2 M hydrochloric acid when the reaction is complete. Filter off the solid obtained and recrystallize from water, aqueous ethanol or ethanol.

**(b) Picrate**

$$\begin{array}{c}CO_2H\\|\\R\cdot NH_2\end{array} + 2,4,6\text{-}(NO_2)_3C_6H_2\cdot OH \longrightarrow \left[\begin{array}{c}CO_2H\\|\\R\cdot NH_3\end{array}\right]^+ [2,4,6\text{-}(NO_2)_3C_6H_2O]^-$$

Prepare as described under *Amines, Primary and Secondary* (p. 29).

**(c) Ethanoyl (acetyl) derivative**

$$\begin{array}{c}CO_2H\\|\\R\cdot NH_2\end{array} + (CH_3\cdot CO)_2O \longrightarrow \begin{array}{c}CO_2H\\|\\R\cdot NH\cdot CO\cdot CH_3\end{array} + CH_3\cdot CO_2H$$

Prepare as described under *Amines, Primary and Secondary* (p. 29).

## CARBOHYDRATES

**(a) β-Ethanoate (β-acetate), e.g. of D-glucose:**

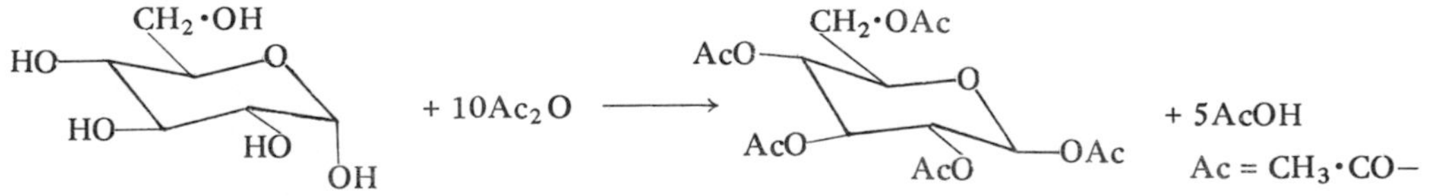

Treat the carbohydrate (0.5 g) with anhydrous sodium ethanoate (acetate) (0·5 g) and ethanoic (acetic) anhydride (3 cm$^3$). Heat on a boiling-water bath for 90 min and pour the product into water (25 cm$^3$). Filter off the solid obtained after stirring, wash well with water and recrystallize from ethanol. If an oil is obtained, decant the water and induce crystallization by scratching with a glass rod.

**(b) Benzoate** (of glucose and fructose only), e.g. of D-glucose:

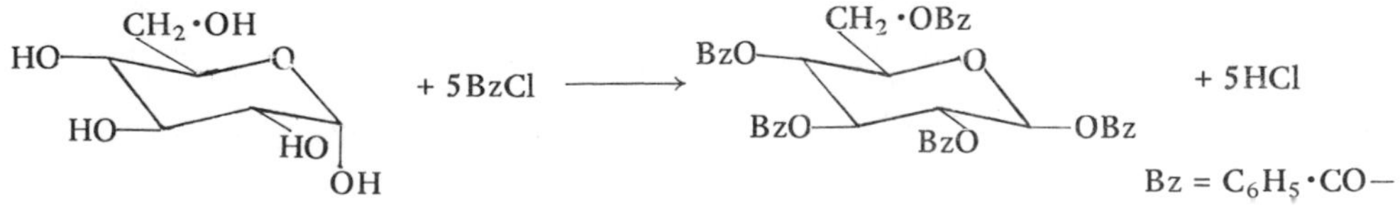

Prepare as described under *Alcohols* (p. 26).

**(c) 4-N-Glycosylaminobenzoic acid**, e.g. of D-glucose:

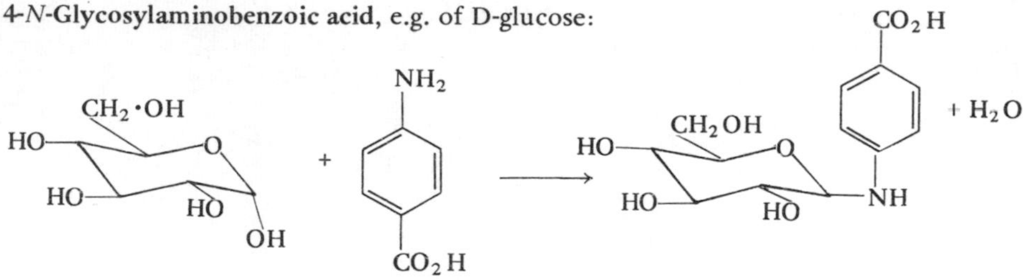

Heat the carbohydrate (1 g) with water (not more than 0.5 cm$^3$) on a boiling-water bath. When most of the sugar has dissolved, add 4-aminobenzoic acid (1 g) in three portions. Continue heating for not more than 2 min. Remove the reaction mixture from the water bath and add methanol (4 ml). Cool in ice if necessary and filter off the solid. Wash it with a small volume of cold methanol and dry at room temperature in air or under vacuum. The product may be recrystallized from ethanol if desired. The melting point of this derivative is best determined by introducing the sample into the apparatus which has been pre-heated to 120°, or to 170° if an approximate determination of melting point indicated that the compound melted above 180°.

**(d) Osazone**

$$\begin{array}{c} CHO \\ | \\ CH\cdot OH + 2\ C_6H_5\cdot NH\cdot NH_2 \\ | \\ R \end{array} \longrightarrow \begin{array}{c} CH=N\cdot NH\cdot C_6H_5 \\ | \\ C=N\cdot NH\cdot C_6H_5 \quad + 2\ H_2O \\ | \\ R \end{array}$$

Dissolve the carbohydrate (1 g) in water (5 cm$^3$) and add phenylhydrazine (1 cm$^3$) and ethanoic (acetic) acid (1 cm$^3$). Heat the mixture on a boiling-water bath for 30 min and allow to cool. Filter off the product, wash well with cold water and recrystallize from ethanol.

# CARBOXYLIC ACIDS

**(a) Amide, anilide and 4-toluidide**

$$RCO_2H \xrightarrow{SOCl_2} RCOCl \begin{cases} \xrightarrow{NH_3} RCO\cdot NH_2 \\ \xrightarrow{C_6H_5\cdot NH_2} RCO\cdot NH\cdot C_6H_5 \\ \xrightarrow{4-CH_3\cdot C_6H_4\cdot NH_2} RCO\cdot NH\cdot C_6H_4\cdot 4-CH_3 \end{cases}$$

To the acid (1 g), add thionyl chloride (2 cm$^3$) and dimethylmethanamide (dimethylformamide, 2 drops) and heat under reflux on a boiling-water bath until no further reaction occurs (about 30 min). Distil off the excess of thionyl chloride and add the residue to an excess of ammonia (0.880, 10 cm$^3$), or of aniline (1 cm$^3$) or 4-toluidine (1 g) in benzene (10 cm$^3$). Moisture must be excluded until the base has been added. Filter off the solid product and recrystallize from water, aqueous ethanol or ethanol.

**(b) 4-Bromophenacyl ester and 4-phenylphenacyl ester**

$$RCO_2H + ArCO\cdot CH_2Br \longrightarrow RCO\cdot O\cdot CH_2\cdot COAr + HBr$$

where Ar = $4-BrC_6H_4-$ or $4-C_6H_5\cdot C_6H_4-$.

Prepare a solution of the acid (1 g) in an equivalent amount of sodium hydroxide solution, make *slightly* acid to litmus by adding a few drops of 2 M hydrochloric acid and add the phenacyl bromide (1 g) (CAUTION: these bromides are eye and skin irritants) in ethanolic solution. Heat to boiling, adding more ethanol if solution is not complete. Continue refluxing for 1,2 or 3 hours depending on whether the acid is mono-, di- or tri-basic. Cool and filter the product. Recrystallize from ethanol, aqueous ethanol or benzene.

# ENOLS

## Semicarbazone and 2,4-dinitrophenylhydrazone

$$RC:CHR' \rightleftharpoons RC \cdot CH_2 R' \xrightarrow{H_2N \cdot NH \cdot CO \cdot NH_2} RC \cdot CH_2 R' + H_2O$$

with $OH$, $O$ below the left side, and $N \cdot NH \cdot CO \cdot NH_2$ below the product.

$$\downarrow 2,4\text{-}(NO_2)_2 \cdot C_6 H_3 \cdot NH \cdot NH_2$$

$$RC \cdot CH_2 R'$$

$$2,4\text{-}(NO_2)_2 C_6 H_3 \cdot NH \cdot N$$

Prepare as described under *Aldehydes* (p. 27).

# ESTERS

## (a) Hydrolysis

$$RCO_2 R' + H_2O \xrightarrow{HO^-} RCO_2 H + R'OH$$

Various methods are available for the hydrolysis of esters to the parent acid and alcohol or phenol depending on the ease of hydrolysis of the ester and the boiling point of R'OH. Three methods are described below which make use of potassium hydroxide in different solvents, viz., water, ethanol and di(2-hydroxyethyl) ether (diethylene glycol).

*Aqueous alkali* — Reflux the ester (5 g) with 30% aqueous potassium hydroxide (40 cm$^3$) until hydrolysis is complete, normally indicated by a change in appearance or odour of the reaction mixture. If a solid separates (usually a sparingly soluble salt of the acid component), filter, wash well with water and characterize as the 4-bromo- or 4-phenylphenacyl ester as described under *Carboxylic Acids* (p. 32). If a liquid separates, this may be an insoluble alcohol. In such cases, extract with ether, dry the ether extract over anhydrous sodium sulphate and evaporate the ether on a water bath. Characterize the compound obtained as described under *Alcohols* (p. 26).

Where a homogeneous solution is obtained, saturate with solid potassium carbonate and extract with ether. Dry the ether extract over anhydrous sodium sulphate and evaporate the ether. Characterize the residue as described under *Alcohols* (p. 26). Acidify the aqueous solution with hydrochloric acid and filter off any solid product. Wash well with water and characterize as described under either *Carboxylic Acids* (p. 32) or *Phenols* (p. 41), see Note 1. If no solid was obtained, saturate with calcium chloride and extract with ether. Evaporate the ether and characterize the residue as described under *Carboxylic Acids* (p. 32) or *Phenols* (p. 41), see Note 1.

*Methanolic alkali* — (for esters of high-boiling alcohols and phenols). Reflux the ester (5 g) with 20% methanolic potassium hydroxide (40 cm$^3$) until hydrolysis is complete. If a solid separates, filter off, wash well with methanol and characterize as described under *Carboxylic Acids* (p. 32) or *Phenols* (p. 41), see Note 1. Carefully distil the excess methanol from the filtrate and characterize the residue as described under *Alcohols* (p. 26) or *Phenols* (p. 41). If a homogeneous solution is obtained after hydrolysis, distil off the bulk of the methanol on a water bath, cool and extract the residue with ether. Dry the ether extract over anhydrous sodium sulphate, evaporate the ether and characterize the residue as described under *Alcohols* (p. 26). Characterize the ether-insoluble residue remaining as described under *Carboxylic Acids* (p. 32).

*Alkali in diethylene glycol* — (for esters resistant to hydrolysis). To the ester (5 g) add a solution consisting of potassium hydroxide (2 g) in diethylene glycol (10 cm$^3$) and water (2 cm$^3$). Reflux the mixture for 5 min. Distil off all the volatile material (water and alcohol) and saturate this distillate with potassium carbonate before extracting with ether. Dry the ether extract, distil off the solvent using a fractionating column and characterize the residue as described under *Alcohols* (p. 26). If no alcohol is detected, the residue remaining after ether extraction will contain a carboxylic acid and phenol. Dissolve in water and acidify with hydrochloric acid. Filter off any solid obtained and characterize as described under *Carboxylic Acids* (p. 32) or *Phenols* (p. 41), see Note 1. The filtrate will contain the remaining component which should be characterized in the same manner.

*Note 1.* Where carboxylic acids and phenols of similar solubility in the solvents used are produced together, they may be separated as follows: acidify the residue and saturate the solution with calcium chloride. Extract with ether and treat the ether extract with 5% sodium bicarbonate solution. Separate the ether layer and dry over anhydrous sodium sulphate. Evaporate the ether and characterize the residue as described under *Phenols* (p. 41). Acidify the bicarbonate extract, saturate with calcium chloride and extract with ether. Dry the ether extract over anhydrous sodium sulphate and evaporate the ether. Characterize the residue as described under *Carboxylic Acids* (p. 32).

# ETHERS

(a) **Nitro derivative**

where $n = 1, 2$ or $3$

Nitration is sometimes a hazardous procedure and must always be undertaken with caution. Various nitration methods are available depending on the ease with which the particular compound may be nitrated. These methods are described below and the preferred method is indicated in the melting point tables.

(i) Mix equal volumes of concentrated sulphuric acid and concentrated nitric acid (5 cm$^3$) and add the organic compound (0·5 g). Maintain the temperature at 25° by cooling and shaking until the reaction is complete. If no reaction occurs, the mixture should be heated cautiously to start the reaction. Pour the resulting material into water (50 cm$^3$) and stir. Filter off the solid which is formed.

(ii) Mix concentrated sulphuric acid (5 cm$^3$) and fuming nitric acid (3 cm$^3$) (CAUTION) and cool to room temperature. Add the organic compound (0.5 g) with continuous shaking and cooling. When the initial reaction has subsided, heat for 5 min on a boiling-water bath. Pour into cold water (50 cm$^3$) and filter off the solid product. Sometimes an oil is obtained, but this will usually solidify on vigorous stirring and scratching.

(iii) Treat the organic compound (0.5 g) dropwise with fuming nitric acid (3 cm$^3$) with continuous cooling in ice. When the reaction subsides, allow the mixture to stand at room temperature for 5 min before pouring it into water (50 cm$^3$). Filter off the solid obtained.

(iv) Dissolve the organic compound (0.5 g) in the minimum of glacial acetic acid and add a mixture of fuming nitric acid (2 cm$^3$) and ethanoic (acetic) acid (2 cm$^3$). Heat the mixture to boiling and allow to stand until cold. Pour into water (50 cm$^3$) and filter off the resulting solid.

(v) As described in (iv) above but keep the mixture at 20$^\circ$ by cooling in ice. After standing for 5 min, dilute with water and filter off the solid product.

In all the above cases, the crude product should be thoroughly washed with water and recrystallized from aqueous ethanol, ethanol or benzene.

### (b) Alkyl 3,5-dinitrobenzoate

$$ROR + 3,5\text{-}(NO_2)_2 C_6 H_3 \cdot COCl \longrightarrow 3,5\text{-}(NO_2)_2 C_6 H_3 \cdot CO \cdot OR + RCl$$

Treat the alcohol-free ether (1 g) with powdered anhydrous zinc chloride (0.1 g) and 3,5-dinitrobenzoyl chloride (0.5 g). Reflux gently for 1 hr, pour the product into saturated aqueous sodium carbonate solution (10 cm$^3$) and heat on a boiling-water bath for 1 min. Allow to cool and filter off the solid obtained. Wash with sodium carbonate solution and then water. Dry the solid and extract it with boiling carbon tetrachloride. Evaporate the excess of solvent and allow the derivative to crystallize.

### (c) Picric acid complex

$$ArOR + 2,4,6\text{-}(NO_2)_3 C_6 H_2 \cdot OH \longrightarrow [ArOR]\,[2,4,6\text{-}(NO_2)_3 C_6 H_2 \cdot OH]$$

Prepare as described under *Amines, Primary and Secondary* (p. 29).

### (d) Sulphonamide

Prepare as described under *Halides, Aryl* (p. 37).

### (e) Bromination

where $n$ = 1, 2 or 3.

35

Suspend or dissolve the ether (1 g) in ethanoic (acetic) acid, trichloromethane (chloroform) or carbon tetrachloride (5 cm$^3$) and add dropwise a solution of bromine in the same solvent until the colour of the bromine persists. Allow to stand for up to 15 min, adding more bromine solution if the colour fades. Evaporate the solvent (when using ethanoic (acetic) acid, pour into water) and recrystallize the product from ethanol.

# HALIDES, ALKYL MONO-

## (a) Thiouronium picrate

$$AlkX + H_2N \cdot CS \cdot NH_2 \longrightarrow \left[ \begin{array}{c} NH_2 \\ \| \\ AlkS \cdot C \cdot NH_2 \end{array} \right]^+ X^- \xrightarrow{2,4,6\text{-}(NO_2)_3C_6H_2 \cdot OH}$$

$$\left[ \begin{array}{c} NH_2 \\ \| \\ AlkS \cdot C \cdot NH_2 \end{array} \right]^+ [2,4,6\text{-}(NO_2)_3C_6H_2O]^- + HX$$

Treat the halide (1 cm$^3$) with a solution of thiourea (1.5 g) in water (4 cm$^3$) and ethanol (3 cm$^3$). Heat on a boiling-water bath until solution is complete and then for a further 15 min. Pour the resulting solution into an excess of 1% aqueous picric acid solution and filter off the precipitate which forms. Recrystallize from aqueous ethanol.

## (b) 2-Naphthyl ether

$$AlkX + 2\text{-}C_{10}H_7O^- \longrightarrow 2\text{-}C_{10}H_7 \cdot OAlk + X^-$$

Prepare a mixture of the halide (1 g), potassium hydroxide (1 g) and 2-naphthol (2 g) in ethanol (10 cm$^3$). Boil under reflux for 15 min and add a further portion of potassium hydroxide (2 g) in water (20 cm$^3$). Shake until a solid product is obtained, filter off and wash well with water. Recrystallize from aqueous ethanol. Isolate liquid products by ether extraction.

## (c) 2-Naphthyl ether picrate

$$2\text{-}C_{10}H_7 \cdot OAlk + 2,4,6\text{-}(NO_2)_3C_6H_2 \cdot OH \longrightarrow [2\text{-}C_{10}H_7 \cdot OAlk][2,4,6\text{-}(NO_2)_3C_6H_2 \cdot OH]$$

To the product (0.5 g) from (b) above in ethanol (2 cm$^3$) add cold saturated alcoholic picric acid solution (5 cm$^3$). Collect the precipitate formed and recrystallize from ethanol.

## (d) Oxidation of substituted benzyl halides to the corresponding carboxylic acids

$$ArCH_2X \xrightarrow{[O]} ArCO_2H$$

### (i) Chromic acid oxidation

Carry out as described under *Alcohols* (p. 26).

### (ii) Potassium permanganate oxidation

Mix the halide (1.5 g) with sodium hydroxide (1 g) and potassium permanganate (9 g) in water (100 cm$^3$). Reflux the solution until the colour of the permanganate is discharged and then filter off the manganese dioxide formed. Acidify the filtrate with concentrated hydrochloric acid and filter off the solid obtained. Recrystallize from water, aqueous ethanol or ethanol.

## HALIDES, ALKYL POLY-

**(a) Thiouronium picrate, e.g.**

$$ClCH_2 \cdot CH_2 Cl + 2H_2N \cdot CS \cdot NH_2 \xrightarrow{\text{picric acid}}$$

$$\left[ \begin{array}{c} H_2N \cdot \underset{\underset{NH_2}{\|}}{C} \cdot S \cdot CH_2 \cdot CH_2 \cdot S \cdot \underset{\underset{NH_2}{\|}}{C} \cdot NH_2 \end{array} \right]^{2+} 2[2,4,6\text{-}(NO_2)_3 C_6 H_2 O]^-$$

Prepare as described under *Halides, Alkyl Mono-* (p. 36).

**(b) 2-Naphthyl ether, e.g.**

$$CH_2 Cl_2 + 2[2\text{-}C_{10}H_7 O^-] \longrightarrow [2\text{-}C_{10}H_7 O]_2 CH_2 + 2Cl^-$$

Prepare as described under *Halides, Alkyl Mono-* (p. 36).

## HALIDES, ARYL

**(a) Sulphonamide**

$$ArX + \underset{\text{excess}}{ClSO_2 \cdot OH} \longrightarrow 4\text{-}XAr \cdot SO_2 Cl \xrightarrow{NH_3} 4\text{-}XAr \cdot SO_2 \cdot NH_2$$

Prepare a solution of the halide (1 g) in dry trichloromethane (chloroform) (5 cm$^3$), cool in ice and add chlorosulphonic acid (3 cm$^3$). When the evolution of hydrogen chloride slackens, warm and maintain at room temperature for 30 min (50° for 10 min if reaction is slow). Pour the product into crushed ice, separate the trichloromethane layer, dry over anhydrous sodium sulphate and evaporate the trichloromethane on a boiling-water bath. Add to the residue ammonia solution (0.88, 10 cm$^3$), boil for 10 min (fume cupboard), cool and dilute with water (10 cm$^3$). Filter off the crude sulphonamide and recrystallize from aqueous ethanol.

**(b) Nitro-derivative**

$$ArX \xrightarrow{NO_2^+} 4\text{-}XAr \cdot NO_2$$

Prepare as described under *Ethers* (p. 34).

**(c) Picric acid complex**

$$ArX + 2,4,6\text{-}(NO_2)_3 C_6 H_2 \cdot OH \longrightarrow [ArX][2,4,6\text{-}(NO_2)_3 C_6 H_2 \cdot OH]$$

Prepare as described under *Amines, Primary and Secondary* (picrate) (p. 29).

## HYDRAZINE DERIVATIVES

**(a) Benzoyl derivative**

$$RNH \cdot NH_2 + C_6 H_5 COCl \longrightarrow RNH \cdot NH \cdot CO \cdot C_6 H_5 + HCl$$

Prepare as described under *Amines, Primary and Secondary* (p. 29).

**(b) Hydrazone from hydrazine derivative**

$$RR'N \cdot NH_2 + C_6H_5 \cdot CHO \longrightarrow RR'N \cdot N{:}CH \cdot C_6H_5 + H_2O$$

Dissolve the hydrazine derivative (0.5 g) in ethanoic (acetic) acid (1 cm$^3$) and water (1 cm$^3$). To this solution add benzaldehyde (0.5 g) and warm for 5 min. Add water (5 cm$^3$) and filter the solid obtained. Recrystallize from ethanol or ethanoic acid.

# HYDROCARBONS

**(a) Diels–Alder adduct with maleic anhydride or benzoquinone, e.g.**

Dissolve the hydrocarbon (1 g) in xylene (5–10 cm$^3$) and add powdered maleic anhydride (1 g) or benzoquinone (1 g). Reflux the mixture for 25 min and cool. If no solid separates, add small quantities of light petrolum (60–80°) to precipitate the product. Filter off the solid obtained and wash with a little light petroleum before recrystallizing from methanol, cyclohexane or xylene.

**(b) Sulphonamide**

$$ArH + \underset{\text{excess}}{ClSO_2OH} \longrightarrow ArSO_2Cl \xrightarrow{NH_3} ArSO_2 \cdot NH_2$$

Prepare as describe under *Halides, Aryl* (p. 37).

**(c) Mercury derivative of alkynes**

$$2RC{:}CH + K_2HgI_4 \longrightarrow (RC{:}C)_2Hg$$

Dissolve mercuric chloride (6.6 g) in a solution of potassium iodide (16.3 g) in water (16.3 cm$^3$) and add 2 M sodium hydroxide (12.5 cm$^3$). Dissolve the alkyne (0.5 g) in ethanol (10 cm$^3$) and add it dropwise to the prepared solution (10 cm$^3$); filter off the precipitate immediately and wash with 50% aqueous ethanol. Recrystallize the product from ethanol or benzene.

**(d) Picric acid and styphnic acid derivatives**

Prepare as described under *Amines, Primary and Secondary* (picrate) (p. 29).

# KETONES

### (a) 2,4-Dinitrophenylhydrazone

$$RR'C{:}O + 2,4\text{-}(NO_2)_2 C_6 H_3 {\cdot}NH{\cdot}NH_2 \longrightarrow 2,4\text{-}(NO_2)_2 C_6 H_3 {\cdot}NH{\cdot}N{:}CRR' + H_2 O$$

Prepare as described under *Aldehydes* (p. 27).

### (b) Semicarbazone

$$RR'C{:}O + H_2 N{\cdot}NH{\cdot}CO{\cdot}NH_2 \longrightarrow RR'C{:}N{\cdot}NH{\cdot}CO{\cdot}NH_2 + H_2 O$$

Prepare as described under *Aldehydes* (p. 27).

### (c) Oxime

$$RR'C{:}O + H_2 N{\cdot}OH \longrightarrow RR'C{:}N{\cdot}OH + H_2 O$$

Dissolve the ketone (0.5 g) in ethanol (3 cm³) and water (1 cm³) and add hydroxylamine hydrochloride (0.3 g) followed by sodium hydroxide (0.5 g). When solution is complete, reflux for 5–10 min. Cool in ice and acidify with hydrochloric acid (use litmus paper). Filter off the product and recrystallize from ethanol.

### (d) 4-Nitrophenylhydrazone and phenylhydrazone

$$RR'C{:}O + ArNH{\cdot}NH_2 \longrightarrow RR'C{:}N{\cdot}NHAr + H_2 O$$

where Ar = $4\text{-}NO_2 {\cdot}C_6 H_4 -$ or $C_6 H_5 -$.

Prepare as described under *Aldehydes* (p. 27).

### (e) Benzylidene derivative

$$\begin{array}{c} RCH_2 \\ | \\ C{:}O + 2C_6 H_5 {\cdot}CHO \\ | \\ R'CH_2 \end{array} \longrightarrow \begin{array}{c} RC{:}CHC_6 H_6 \\ | \\ C{:}O \\ | \\ R'C{:}CH{\cdot}C_6 H_5 \end{array} + 2H_2 O$$

Shake a mixture of the ketone (0.5 g), benzaldehyde (1.2 cm³) in a little ethanol and 4 M sodium hydroxide (0.5 cm³). Allow to stand at room temperature until a crystalline product is obtained. Scratching the vessel with a glass rod will often induce crystallization. Filter off the product and recrystallize from ethanol.

# NITRILES

### (a) Amide

$$RCN \xrightarrow{H_2 O_2} RCO{\cdot}NH_2$$

Prepare a solution containing 20 volume hydrogen peroxide (10 cm³) and 2 M sodium hydroxide (2 cm³), add the nitrile (0.5 g) and heat to 40° on a water bath. Shake frequently and finally filter off the solid product obtained. Wash and recrystallize from water, aqueous ethanol or ethanol.

(b) **Nitro-derivative** (for aromatic nitriles only)

$$ArCN \xrightarrow{NO_2^+} 3\text{-}NO_2 \cdot Ar \cdot CN$$

Prepare as described under *Ethers*, Method (i) (p. 34).

(c) **Hydrolysis to carboxylic acid**

$$RCN \longrightarrow RCO_2H$$

Reflux the nitrile (1 g) with either 8 M sodium hydroxide (5 cm$^3$) for aliphatic nitriles or 7 M sulphuric acid (5 cm$^3$) for aromatic nitriles for 1 hr. Cool the resulting solution and add excess hydrochloric acid (aliphatic nitriles) or water (aromatic nitriles). Characterize the aliphatic acid produced as its 4-bromophenacyl ester as described under *Carboxylic Acids* (p. 32).

Aromatic acids produced may be filtered off and recrystallized from water, aqueous ethanol or ethanol.

# NITRO-, HALOGENONITRO-COMPOUNDS AND NITRO-ETHERS

(a) **Nitration**

Prepare as described under *Ethers*, (p. 34).

(b) **Benzylidene derivative**

$$RCH_2 \cdot NO_2 + C_6H_5 \cdot CHO \longrightarrow \underset{\overset{\|}{CH \cdot C_6H_5}}{RC \cdot NO_2} + H_2O$$

Prepare as described under *Ketones*, (p. 39).

(c) **Oxidation of alkyl side-chain to carboxylic acid**

$$ArCH_3 \xrightarrow{[O]} ArCO_2H$$

Prepare as described under *Alcohols*, (p. 26).

(d) **Reduction to amine**

$$RNO_2 \xrightarrow{[H]} RNH_2$$

Suspend the nitro-compound (1 g) in concentrated hydrochloric acid (10 cm$^3$) and add ethanol (2 cm$^3$) and tin (3 g). Cool until the initial reaction subsides and then heat under reflux for 30 min. Filter the solution, cool the filtrate and basify with 5 M sodium hydroxide, adding sufficient alkali to dissolve the precipitate of stannous hydroxide formed. Extract the free amine with ether, dry the ether extract over anhydrous sodium sulphate and evaporate the ether (CARE). Further conversion to crystalline derivatives should be carried out as described under *Amines, Primary and Secondary*, (p. 29).

(e) **Partial reduction, e.g.**

$$1,3\text{-}(NO_2)_2 C_6 H_4 \longrightarrow 3\text{-}NO_2 \cdot C_6 H_4 \cdot NH_2$$

Dissolve the nitro-compound (1 g) in ethanol (10 cm$^3$) and add ammonia solution (0.88, 1 cm$^3$). Saturate the cold solution with hydrogen sulphide and reflux on a boiling-water bath for 30 min. Cool, resaturate with hydrogen sulphide and reflux for a further 30 min. Pour into cold water and filter the solid obtained. Extract the solid with 2 M hydrochloric acid, basify this extract with ammonia solution (0.88) and filter the resulting nitro-amine. Recrystallize from aqueous ethanol, ethanol or benzene.

# PHENOLS

(a) **Ethanoate (acetate)**

$$ArOH + CH_3 \cdot COCl \longrightarrow ArO \cdot CO \cdot CH_3 + HCl$$

(i) Dissolve the phenol (0.5 g) in dry pyridine (0.5 cm$^3$) and add ethanoyl (acetyl) chloride (0.5 cm$^3$) dropwise. Shake well after each addition and cool if the temperature rises rapidly. When the addition of ethanoyl chloride is complete, heat to 50—60° for 5 min. Cool, pour into water (15 cm$^3$) and stir until a solid is obtained. Filter off the solid and recrystallize from ethanol or aqueous ethanol.

(ii) Stir the phenol (0.5 g) with ethanoic (acetic) anhydride (0.5 cm$^3$) containing concentrated sulphuric acid (3 drops) at 60° for 15 min. Cool, add water (7 cm$^3$), stir well, and recrystallize the product.

(b) **Benzoate and toluene-4-sulphonate**

$$ArOH + C_6 H_5 \cdot COCl \longrightarrow ArO \cdot CO \cdot C_6 H_5 + HCl$$

$$ArOH + 4\text{-}CH_3 \cdot C_6 H_4 \cdot SO_2 Cl \longrightarrow 4\text{-}CH_3 \cdot C_6 H_4 \cdot SO_2 \cdot OAr + HCl$$

Prepare as described under *Alcohols* (p. 26). For nitrophenols it is preferable to replace this Schotten—Baumann method by one in which pyridine is the base, as described under *Alcohols*, method c (p. 26).

(c) **4-Nitrobenzoate and 3,5-dinitrobenzoate**

$$ArOH + Ar'COCl \longrightarrow ArO \cdot COAr' + HCl$$

where Ar$'$ = 4-NO$_2 \cdot$C$_6$H$_4^-$ or 3,5-(NO$_2$)$_2$C$_6$H$_3^-$.

Prepare as described under *Alcohols*, method c (p. 26).

(d) **Aryloxyethanoic acid**

$$ArOH + ClCH_2 \cdot CO_2 H \longrightarrow ArO \cdot CH_2 \cdot CO_2 H + HCl$$

To a solution of the phenol (0.5 g) in 5 M sodium hydroxide (3 cm$^3$) add chloroethanoic (chloroacetic) acid (0.5 g) (CAUTION: this acid must not be allowed to come into contact with the skin). Add a little water if any solid is formed in the hot solution. Heat on a boiling-water bath for 1 hr, cool, acidify with 2 M hydrochloric acid to Congo Red and extract with ether.

Extract the ethereal layer with 2 M sodium carbonate solution. If the sodium salt of the acid separates, remove by filtration and treat the solid with 2 M hydrochloric acid. The resulting solid is the required derivative. If no solid separates, acidify the sodium carbonate extract with 2 M hydrochloric acid. Filter off the solid obtained. Recrystallize the product from water, aqueous ethanol or ethanol.

**(e)  Derivative for picric and styphnic acids, e.g.**

$$2,4,6\text{-}(NO_2)_3 C_6 H_2 \cdot OH + C_{10}H_8 \longrightarrow [2,4,6\text{-}(NO_2)_3 C_6 H_2 \cdot OH][C_{10}H_8]$$

Prepare saturated solutions of either picric or styphnic acid and naphthalene in ethanol and mix. Warm gently for a few minutes and cool. A crystalline product is readily obtained which may be recrystallized from ethanol.

# QUINONES

**(a)  Oxime, e.g.**

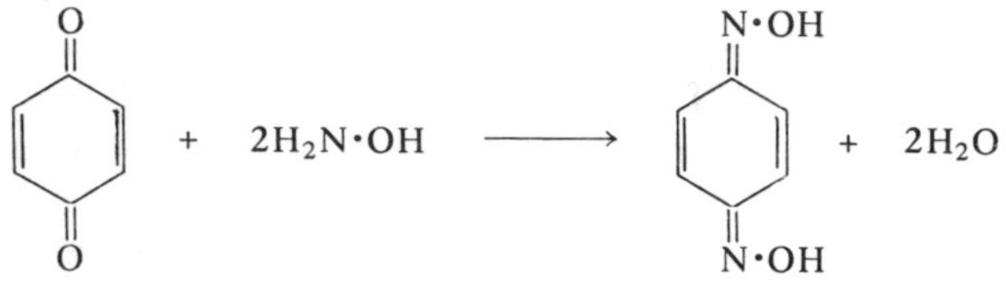

Prepare as described under *Ketones*, (p. 39).

**(b)  Semicarbazone, e.g.**

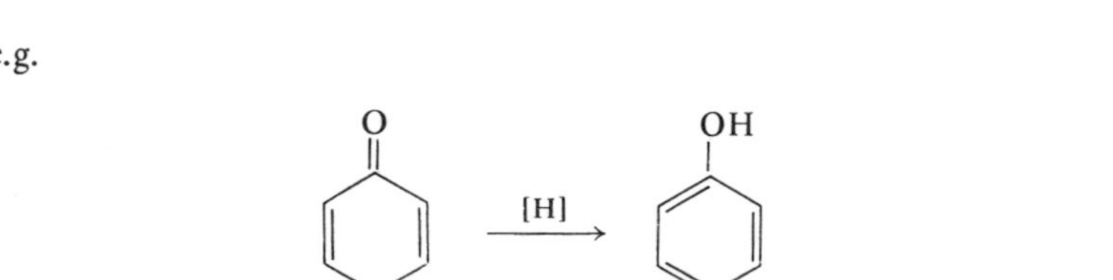

Prepare as described under *Aldehydes*, (p. 27).

**(c)  Quinol, e.g.**

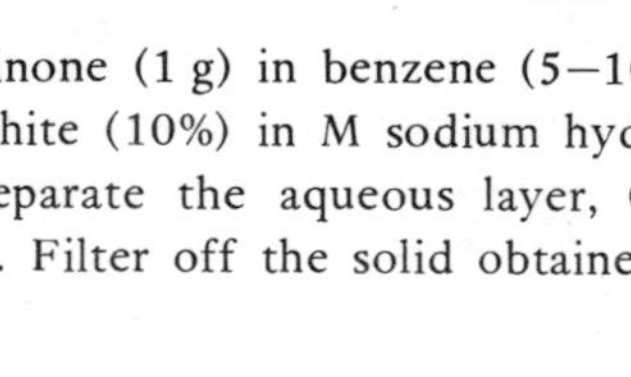

Dissolve or suspend the quinone (1 g) in benzene (5–10 cm$^3$) and treat with a solution (20 cm$^3$) of sodium hydrosulphite (10%) in M sodium hydroxide. Shake until the quinone colour has disappeared and separate the aqueous layer, Cool this (ice) and acidify with concentrated hydrochloric acid. Filter off the solid obtained and recrystallize from water or ethanol.

# SULPHONIC ACIDS AND THEIR DERIVATIVES

### (a) Amide

$$RSO_2 \cdot OH + PCl_5 \longrightarrow RSO_2Cl \xrightarrow{NH_3} RSO_2 \cdot NH_2$$

Mix the dry sulphonic acid (1 g) or the dry sodium salt (1 g) with phosphorus pentachloride (2 g) and heat on a boiling-water bath, taking care to exclude water vapour from the reaction vessel. When reaction has ceased, add water (15 cm$^3$) and stir. Decant the water and add ammonia solution (0.88, 3 cm$^3$) to the residue. Heat on a boiling-water bath for 5–10 min and cool. Filter off the solid product, wash well with water and recrystallize from water or aqueous ethanol.

### (b) Anilide

$$RSO_2 \cdot OH + PCl_5 \longrightarrow RSO_2Cl \xrightarrow{C_6H_5 \cdot NH_2} RSO_2 \cdot NH \cdot C_6H_5$$

Use the method described above but replace the ammonia solution with aniline (1 cm$^3$).

### (c) S-Benzylisothiouronium salt

$$RSO_2 \cdot O^-Na^+ + \left[ \begin{array}{c} C_6H_5 \cdot CH_2 \cdot S \cdot C{:}NH_2 \\ | \\ NH_2 \end{array} \right]^+ Cl^- \longrightarrow$$

$$\left[ \begin{array}{c} C_6H_5 \cdot CH_2 \cdot S \cdot C{:}NH_2 \\ | \\ NH_2 \end{array} \right]^+ [O \cdot SO_2R]^- + NaCl$$

Prepare the sodium salt of the sulphonic acid (0.5 g) in water (3 cm$^3$) by addition of 2 M sodium hydroxide until the solution is just alkaline to phenolphthalein. Neutralize the excess of alkali with a further addition of the sulphonic acid (or 2 M hydrochloric acid) and add a solution of S-benzylthiouronium chloride (2 g) in water (5 cm$^3$). Cool in ice, filter off the crystalline product and recrystallize from water, aqueous ethanol or ethanol.

### (d) Xanthyl derivative of sulphonamides

Treat the sulphonamide (0.5 g) with xanthydrol (0.5 g) in ethanoic (acetic) acid (25 cm$^3$) and heat the mixture until solution is complete. Allow to stand at room temperature until a solid separates. Water may be added if no solid separates. Filter off the product and recrystallize from aqueous dioxan.

### (e) Benzoyl derivative of sulphonamides

$$RSO_2 \cdot NH_2 + C_6H_5COCl \longrightarrow RSO_2 \cdot NH \cdot CO \cdot C_6H_5 + HCl$$

Prepare as described under *Alcohols*, (p. 26).

### (f) Ethanoyl (acetyl) derivative of sulphonamides

$$RSO_2 \cdot NH_2 + CH_3 \cdot COCl \longrightarrow RSO_2 \cdot NH \cdot CO \cdot CH_3 + HCl$$

To the sulphonamide (1 g) add ethanoyl (acetyl) chloride (3 cm$^3$) and reflux for 30 min, adding ethanoic (acetic) acid (up to 2 cm$^3$) if solution is not complete. Remove the excess of ethanoyl chloride by vacuum distillation and pour the residue into ice cold water (25 cm$^3$). Stir the product until is solidifies, filter off, wash well with water and recrystallize from aqueous ethanol.

## THIOETHERS (SULPHIDES)

### (a) Sulphone

$$RSR' \xrightarrow{[O]} RSO_2R'$$

Dissolve the thioether (1 g) in the minimum of ethanoic (acetic) acid and add 3% potassium permanganate solution as long as the colour is discharged. If starting material is precipitated during this addition, more ethanoic acid should be added. When reaction is complete, pass in sulphur dioxide until the manganese dioxide precipitate has just dissolved. Add crushed ice and filter off the solid sulphone. Wash well with water and recrystallize from ethanol.

## THIOLS AND THIOPHENOLS

### (a) 2,4-Dinitrophenyl|sulphide

$$RSH + 2,4\text{-}(NO_2)_2C_6H_3Cl \longrightarrow 2,4\text{-}(NO_2)_2C_6H_3 \cdot SR + HCl$$

Dissolve the thiol (1 g) in ethanol (30 cm$^3$) and add sodium hydroxide (0·4 g) in ethanol (3 cm$^3$) followed by 2,4-dinitrochlorobenzene (2 g) (CAUTION − skin irritant) in ethanol (10 cm$^3$). Reflux on a boiling-water bath for 10 min, filter and allow the filtrate to cool. Recrystallize the product from ethanol.

### (b) Hydrogen 3-nitrophthaloyl derivative

Prepare as described under *Alcohols* (p. 26).

### (c) 3,5-Dinitrobenzoyl derivative

$$RSH + 3,5\text{-}(NO_2)_2C_6H_3 \cdot COCl \longrightarrow 3,5\text{-}(NO_2)_2C_6H_3 \cdot CO \cdot SR + HCl$$

Prepare as described under *Alcohols* (p. 26).

# 5
## TABLES OF ORGANIC COMPOUNDS
## AND THEIR DERIVATIVES

### EXPLANATORY NOTES ON THE TABLES
### OF COMPOUNDS AND DERIVATIVES

1. In each table the compounds are listed in the order of increasing boiling point if they are liquids or solids melting below 40°. Compounds which melt at 40° or above are divided from the liquids by a horizontal line and are arranged in the order of increasing melting point although the boiling point is sometimes also included.

2. Boiling points are given at atmospheric pressure except for a few high boiling compounds whose boiling points are given at reduced pressure and are written thus: 94/12 mm, which means a boiling point of 94° at 12 mm pressure.

3. Boiling and melting points are given to the nearest whole number and as one value only. This approximation is made for simplicity and because of slight variation in the degree of accuracy of different thermometers and in the personal element in determining the melting or boiling point. For example, a melting point which is recorded in the literature as 172—174° or as 173.5° is given in the tables as 173°.

4. The figures given in the columns of derivatives are melting points. Two different values of a boiling or melting point are recorded in the chemical literature for a few compounds. The second (usually less frequently encountered) value is given in the tables in parentheses below the more common value.

5. For compounds which exist as enantiomorphs, the constants given are those of the racemic or (±)-modification unless otherwise stated.

6. The following abbreviations are used in the tables:

| | |
|---|---|
| *anhyd*. anhydrous | *dil*. dilute |
| *aq*. aqueous | *hyd*. hydrate |
| *conc*. concentrated | *insol*. insoluble |
| *d*. decomposition | *sol*. soluble |
| *deriv*. derivative | *subl*. sublimes |

7. When the colour of a compound is other than white it is given in the last column of the table which also contains supplementary information which may assist in the identification of the compound.

## Table 1. Acetals

| Acetal | B.p. | Aldehyde | 4-Nitrophenyl-hydrazone (p. 28) | 2,4-Dinitro-phenylhydra-zone (p. 27) | Alcohol | K alkyl xanthate (p. 26) |
|---|---|---|---|---|---|---|
| Dimethoxymethane (Methylal) | 45 | Methanol | 182 | 167 | Methanol | 182 |
| 1,1-Dimethoxyethane (Dimethylacetal) | 64 | Ethanol | 128 | 168 | Methanol | 182 |
| Diethoxymethane (Ethylal) | 89 | Methanol | 182 | 167 | Ethanol | 225 |
| 1,1-Diethoxyethane (Acetal) | 102 | Ethanol | 128 | 168 | Ethanol | 225 |
| 1,1-Diethoxyprop-2-ene (Acrolein acetal) | 126 | Propenal (Acrolein) | 151 | 165 | Ethanol | 225 |
| 1,1-Dipropoxymethane | 140 | Methanol | 182 | 167 | Propan-1-ol | 233 |
| 1,1-Dibutoxyethane | 187 | Ethanol | 128 | 168 | Butan-1-ol | 255 |
| αα-Dimethoxytoluene | 198 | Benzaldehyde | 192 | 237 | Methanol | 182 |
| αα-Diethoxytoluene | 222 | Benzaldehyde | 192 | 237 | Ethanol | 225 |

## Table 2. Alcohols (C, H and O)

| | B.p. | M.p. | 3,5-Dinitro-benzoate (p. 26) | H 3-nitro-phthalate (p. 26) | 4-Nitro-benzoate (p. 26) | Notes |
|---|---|---|---|---|---|---|
| Methanol | 65 | | 109 | 153* | 96 | *Anhydrous; monohydrate melts <100 but if dried at 80, it becomes anhydrous |
| Ethanol | 78 | | 94 | 157 | 56 | |
| Propan-2-ol | 83 | | 122 | 154 | 110 | |
| 2-Methylpropan-2-ol (t-Butanol) | 83 | 25 | 142 | | 116 | |
| Propan-1-ol | 97 | | 75 | 144 | 35 | |
| Prop-2-enol (Allyl alcohol) | 97 | | 50 | 124 | 28 | Unsaturated |
| Butan-2-ol (s-Butanol) | 100 | | 76 | 131 | 25 | |
| 2-Methylbutan-2-ol (t-Pentyl alcohol) | 102 | | 118 | | 85 | |
| 2-Methylpropan-1-ol (Isobutanol) | 108 | | 88 | 183 | 69 | |
| 3-Methylbutan-2-ol | 113 | | 76 | 127 | | |
| Pentan-3-ol | 116 | | 100 | 121 | 17 | |
| Butan-1-ol | 117 | | 64 | 147 | 35 | |
| Pentan-2-ol | 120 | | 62 | 103 | 17 | |
| 3-Methylpentan-3-ol | 123 | | 96 | | 69 | |
| 2-Methoxyethanol (Methyl cellosolve) | 125 | | | 129 | 50 | |
| 2-Methylbutan-1-ol (amyl alcohol) | 128 | | 70 | 158 | | |
| 3-Methylbutan-1-ol | 132 | | 62 | 164 | 21 | |
| 4-Methylpentan-2-ol | 132 | | 65 | 166 | 26 | |
| 2-Ethoxyethanol (Ethyl cellosolve) | 135 | | 75 | 121* | | *Anhydrous; monohydrate, 94 |
| Hexan-3-ol | 136 | | 77 | 127 | | |
| Pentan-1-ol | 138 | | 46 | 136 | oil | |
| 2,4-Dimethylpentan-3-ol | 140 | | 38 | 151 | 40 | |
| Cyclopentanol | 140 | | 115 | | 62 | |

**Table 2.** Alcohols (C, H and O) (*cont.*)

| | B.p. | M.p. | 3,5-Dinitro-benzoate (p. 26) | H 3-nitro-phthalate (p. 26) | 4-Nitro-benzoate (p. 26) | Notes |
|---|---|---|---|---|---|---|
| 3-Hydroxybutan-2-one (Acetoin) | 145 | | | | | See Table 26 |
| 1-Hydroxypropan-2-one (Acetol) | 146 | | | | | See Table 26 |
| 2-Methylpentan-1-ol | 148 | | 50 | 145 | | |
| 2-Ethylbutan-1-ol | 149 | | 51 | 147 | | |
| 4-Methylpentan-1-ol | 152 | | 70 | 140 | | |
| -Hexan-1-ol | 156 | | 61 | 124 | | |
| Heptan-2-ol | 158 | | 50 | | | $K_2Cr_2O_7-H_2SO_4$ (p. 27) → heptan-2-one |
| Cyclohexanol | 161 | 25 | 113 | 160 | 52 | |
| Furfuryl alcohol | 170 | | 81 | | 76 | |
| Heptan-1-ol | 176 | | 48 | 127 | oil | |
| Tetrahydrofurfuryl alcohol | 177 | | 83 | | 47 | |
| Octan-2-ol | 179 | | 32 | | 28 | $K_2Cr_2O_7-H_2SO_4$ (p. 27) → octan-2-one |
| 2-Ethylhexan-1-ol | 184 | | | 108 | | |
| Propane-1,2-diol (Propylene glycol) | 187 | | 147 | | 127 | |
| 3,5,5-Trimethylhexan-1-ol | 193 | | 62 | 150 | | |
| Octan-1-ol | 194 | | 62 | 128 | oil | |
| (−)-Linalyl alcohol | 197 | | 135 | | 70 | Unsaturated |
| Ethane-1,2-diol (Ethylene glycol) | 197 | | 169 | | 140 | Ditoluene-4-sulphonate, 93 (p. 26) |
| 1-Phenylethanol | 202 | 20 | 94 | | 43 | |
| Benzyl alcohol | 205 | | 113 | 183 | 85 | |
| Nonan-1-ol | 214 | | 52 | 125 | | |
| Propane-1,3-diol (Trimethylene glycol) | 214 | | 178 | | 119 | Ditoluene-4-sulphonate, 93 (p. 26) |
| Isoborneol | 216 | | 138 | | 129 | |
| 2-Phenylethanol | 219 | | 108 | 123 | 62 | |
| α-Terpineol | 221 | 35 | 78 | | 139 | |
| Tetradecan-1-ol (Myristic alcohol) | 221 | 39 | 67 | 123 | 51 | |
| trans-3,7-Dimethylocta-2,6-dien-1-ol (Geraniol) | 229 | | 63 | 117 | 35 | Unsaturated; Br$_2$ → tetra-bromide, 70 |
| Butane-1,4-diol | 230 | 19 | | | 175 | Dibenzoate, 81; ditoluene-4-sulphonate, 94 (p. 26) |
| Decan-1-ol | 231 | 6 | 57 | 123 | 30 | |
| 2-Phenoxyethanol | 237 (245) | 12 | 105 (74) | 113 | 63 | Toluene-4-sulphonate, 80 (p. 26). |
| 3-Phenylpropan-1-ol | 237 | | 92 | 117 | 46 | |
| Undecan-1-ol | 243 | 15 | 55 | 123 | 30 | |
| Di(2-hydroxyethyl) ether (Diethylene glycol) | 244 | | 150 | | | |
| 3-Phenylprop-2-en-1-ol (Cinnamyl alcohol) | 257 | 33 | 121 | | 78 | Unsaturated; Br$_2$ → dibromide, 74 |
| Dodecan-1-ol (Lauryl alcohol) | 259 | 25 | 60 | 124 | 43 | |
| Propane-1,2,3-triol (Glycerol) | 290d | 18 | | | 188 | |
| (−)-2-Isopropyl-5-methyl-cyclohexanol ((  ) Menthol) | 216 | 42 | 153 | | 61 | |

47

**Table 2.** Alcohols (C, H and O) (*cont.*)

| | B.p. | M.p. | 3,5-Dinitro-benzoate (p. 26) | H 3-nitro-phthalate (p. 26) | 4-Nitro-benzoate (p. 26) | Notes |
|---|---|---|---|---|---|---|
| Hexadecan-1-ol (Cetyl alcohol) | | 50 | 66 | 122 | 52 (58) | |
| Heptadecan-1-ol | | 54 | 121 | 121 | 53 | |
| But-2-yn-1,4-diol | | 55 | 190 | | | Dibenzoate, 76; di-4-tolu-enesulphonate, 94 (p. 26) |
| Octadecan-1-ol (Stearyl alcohol) | | 59 | 66 | 119 | 64 | |
| Diphenylmethanol (Benzhydrol) | | 69 | 141 | | 131 | |
| D-Glucitol (D-Sorbitol) | | 111* | | | | *Anhydrous; hydrate, 90. Hexaethanoate, 99; hexa-benzoate, 129 (p. 26) |
| Benzoin | | 133 | | | 123 | Benzoate, 124 (p. 26). See Table 26 |
| Furoin | | 136 | | | | See Table 26 |
| Triphenylmethanol | | 162 | | | | Ethanoate, 87; benzoate, 162 (p. 26) |
| D-Mannitol | | 166 | | | | Hexaethanoate, 126; hexa-benzoate, 148(129) (p. 26) |
| D-Galactitol (Dulcitol) | | 188 | | | | Hexaethanoate, 171; hexa-benzoate, 188 (p. 26) |
| (+)-Borneol | 212 | 208 | 154 | | 153 | |
| Myo-inositol | | 225 | 86 | | | Hexaethanoate, 212 subl.; hexabenzoate, 258 (p. 26) |
| Pentaerythritol | | 262 (253) | | | | Tetraethanoate, 84; tetra-benzoate, 99 (p. 26) |

**Table 3.** Alcohols (C, H, O and halogen or N)

| | B.p. | M.p. | 3,5-Dinitro-benzoate (p. 26) | H 3-nitro-phthalate (p. 26) | 4-Nitro-benzoate (p. 26) | Notes |
|---|---|---|---|---|---|---|
| 1-Chloropropan-2-ol | 127 | | 77 | | | |
| 2-Chloroethanol (Ethylene chloro-hydrin) | 129 | | 95 (88) | 98 | 56 | |
| 2-Chloropropan-1-ol | 133 | | 76 | | | |
| 2-Dimethylaminoethanol | 135 | | | | | See Table 12 |
| 2-Bromoethanol | 149d | | 85 | 172 | | |
| 2,2,2-Trichloroethanol | 151 | 19 | 142 | | 71 | |
| 2-Diethylaminoethanol | 161 | | | | | See Table 12 |
| 3-Chloropropan-1-ol | 161 | | 77 | | | |
| 1-Aminopropan-2-ol (Isopropanolamine) | 163 | | | | | Picrate, 142; see Table 8 |
| 2-Aminoethanol | 171 | | | | | Picrate, 159; reacts with phthalic anhydride → β-hydroxyethylimide, 127 |
| 3-Nitrobenzyl alcohol | 175/3 mm | 27 | | | | Benzoate, 71; $K_2Cr_2O_7-H_2SO_4$→ acid, 140 (p. 27) |
| 2-Nitrobenzyl alcohol | | 74 | | | | Benzoate, 101; $K_2Cr_2O_7-H_2SO_4$→ acid, 146 (p. 27) |
| 4-Nitrobenzyl alcohol | | 93 | | | | Benzoate, 94; $K_2Cr_2O_7-H_2SO_4$→ acid, 240 (p. 27) |

**Table 4.** Aldehydes (C, H and O)

| | B.p. | M.p. | 2,4-Di-nitro-phenyl-hydra-zone (p. 27) | Semi-carba-zone (p. 27) | Di-meth-one (p. 28) | 4-Nitro-phenyl-hydra-zone (p. 28) | Notes |
|---|---|---|---|---|---|---|---|
| Methanal (Formaldehyde) | −21 | | 167 | 169d | 189 | 182 | 40% aq. solution is formalin |
| Ethanal (Acetaldehyde) | 20 | | 168 | 163 | 140 | 128 | |
| Propanal (Propionaldehyde) | 50 | | 155 | 154* | 155 | 124 | *Recryst. from water |
| Ethanedial (Glyoxal) | 50 | | 327 | 270 | 186* mono | 310d | *Di-, 228 |
| Propenal (Acrolein) | 52 | | 165 | 171 | 192 | 151 | Unsaturated |
| 2-Methylpropanal (Isobutyraldehyde) | 64 | | 182 | 125 | 154 | 131 | |
| 2-Methylprop-2-enal | 73 | | 206 | 198 | | | Unsaturated |
| Butanal | 74 | | 125 | 105 | 136 | 91 | |
| 2,2-Dimethylpropanal (Pivalaldehyde) | 75 | | 209 | 190 | | 119 | |
| 3-Methylbutanal (Isovaleraldehyde) | 92 | | 123 | 132 (107) | 155 | 110 | |
| 2-Methylbutanal | 93 | | 121 | 103 | | | |
| Pentanal (Valeraldehyde) | 103 | | 107 | | 105 | 74 | |
| But-2-enal (Crotonaldehyde) | 103 | | 196 | 200* | 186 | 184 | *Varies with rate of heating. Unsaturated |
| 5-Hydroxymethylfur-furaldehyde | 114 | 35 | 184 | 195d (166) | | 185 | |
| 2-Ethylbutanal | 116 | | 134 | 98 | 102 | | |
| Paraldehyde | 124 | | | | | | Gives ethanal on warming with trace of conc. sulphuric acid |
| Hexanal (Caproaldehyde) | 129 | | 104 | 108 | 109 | 80 | |
| 3-Methylbut-2-enal (β-Methylcrotonalde-hyde) | 135 | | 182 | 221 | | | Unsaturated |
| Tetrahydrofurfural | 144 | | 134 | 166 | 123 | | |
| Heptanal (Heptaldehyde) | 156 | | 108 | 109 | 103 (135) | 73 | |
| Furfural | 161 | | 202* | 203 | 160d | 154 | *Variable |
| Hexahydrobenzaldehyde | 162 | | 172 | 174 | | | |
| 2-Ethylhexanal | 163 | | 121 | 254d | | | |
| Succinaldehyde | 169 | | 280 | | | | Oxime, 172 (p. 28) |
| Octanal | 171 | | 106 | 98 | 90 | 80 | |
| Benzaldehyde | 179 | | 237 | 222* | 195 | 192 | *Varies with rate of heating. Smell of bitter almonds |
| Nonanal | 185 | | 100 | 100 | 86 | | |
| 5-Methylfurfural | 187 | | 212 | 211 | | 130 | |
| Phenylacetaldehyde | 194 | 33 | 121 | 156 | 165 | 151 | |
| 2-Hydroxybenzaldehyde (Salicylaldehyde) | 196 | | 248* | 231 | 211 | 227 | *From ethanol. See Table 30 |
| 3-Tolualdehyde | 199 | | 194 | 224 | 172 | 157 | |
| 2-Tolualdehyde | 200 | | 194 | 212 | 167 | 222 | |
| 4-Tolualdehyde | 204 | | 234 | 234 | | 200 | |
| Decanal | 208 | | 104 | 102 | 92 | | |

49

# Table 4. Aldehydes (C, H and O) (*cont.*)

| | B.p. | M.p. | 2,4-Dinitro-phenyl-hydra-zone (p. 27) | Semi-carba-zone (p. 27) | Di-meth-one (p. 28) | 4-Nitro-phenyl-hydra-zone (p. 28) | Notes |
|---|---|---|---|---|---|---|---|
| Phenoxyacetaldehyde | 215d | | 130 | 145 | | | Oxime, 95 (p. 28) |
| 3-Phenylpropanal | 224 | | 149 | 127 | | 123 | |
| *trans*-3,7-Dimethylocta-2,6-dien-1-al (Geranial, citral a) | 228d | | 116 (108) | 164 | | 195 | Unsaturated |
| 3-Methoxybenzaldehyde | 230 | | 218 | 233d | | 171 | |
| 4-Isopropylbenzaldehyde (Cuminaldehyde) | 235 | | 244 | 211 | 171 | 190 | |
| 2-Methoxybenzaldehyde | 245 | 38 | 253 | 215 | 188 | 205 | |
| 4-Methoxybenzaldehyde (Anisaldehyde) | 248 | | 253d | 210 | 145 | 160 | |
| 3-Phenylpropenal (Cinnamaldehyde) | 252 | | 255d | 215 | 219 (213) | 195 | Unsaturated |
| 3,4-Methylenedioxy-benzaldehyde (Piperonal) | 263 | 37 | 266d | 234 | 178 | 200 | |
| 1-Naphthaldehyde | 292 | 34 | 254 | 221 | | 234 | |
| Tetradecanal (Myristaldehyde) | 155/10 mm | 23 | 108 | 107 | | 95 | |
| Hexadecanal (Palmitaldehyde) | 200/29 mm | 34 | 108 | 108 | | 97 | |
| Octadecanal (Stearaldehyde) | 212/22 mm | 38 | 110 | 108 | | 101 | |
| Dodecanal (Lauraldehyde) | | 44 | 106 | 106 | | 90 | |
| 2,3-Dimethoxybenzal-dehyde | | 54 | 223 (264) | 231 | 150 | | |
| Phthalaldehyde | | 56 | | | | | Phenylhydrazone, 191 (p. 28) |
| 3,4-Dimethoxybenzal-dehyde (Veratraldehyde) | | 58 | 264 | 177 | 173 | | |
| 2-Naphthaldehyde | | 60 | 270 | 245 | | 230 | |
| 2,4-Dimethoxybenzal-dehyde | | 69 | 257 | | | | Oxime, 106 (p. 28) |
| 4-Hydroxy-3-methoxy-benzaldehyde (Vanillin) | | 80 | 271d | 240d | 197 | 227 | Bisulphite addition compound is sol. in water. See Table 30 |
| Phenylglyoxal, hydrate | | 91 | 296 | 217d | | 310 | Dioxime, 168; mono-oxime, 128 (p. 28) |
| 3-Hydroxybenzaldehyde | | 104 | 260 | 199 | | 222 | See Table 30 |
| Terephthalaldehyde | | 116 | | 225d | | 281* | *Sinters at 272. Oxime, 200 (p. 28) |
| 4-Hydroxybenzaldehyde | | 117 | 280d | 222 | 189 | 266 | See Table 30; bisulphite compound is sol. in water |
| 2,4-Dihydroxybenzaldehyde (β-Resorcylaldehyde) | | 135 | 286d (302) | 260d | 226 | | See Table 30 |
| Glyceraldehyde (dimer) | | 142 | 167 | 160d | 197 | | |
| 3,4-Dihydroxybenzaldehyde (Protocatechualdehyde) | | 154 | 275d | 230d | 143d | | See Table 30 |
| 3,5-Dihydroxybenzaldehyde | | 157 | | 223 | | 280* | *Chars without melting |

**Table 5.** Aldehydes (C, H, O and halogen or N)

| | B.p. | M.p. | 2,4-Di-nitro-phenyl-hydra-zone (p. 27) | Semi-carba-zone (p. 27) | Di-meth-one (p. 28) | 4-Nitro-phenyl-hydra-zone (p. 28) | Notes |
|---|---|---|---|---|---|---|---|
| Trichloroethanal (Chloral) | 98 | | 131 | 90d | | | |
| Tribromoethanal (Bromal) | 174* | | | | | | *Anhydrous; hydrate, 54. Oxime, 115 (p. 28) |
| 2-Iodobenzaldehyde | 206 | 37 | 215 | 206 | | | |
| 2-Chlorobenzaldehyde | 208 | 11 | 208 | 225 | 205 | 249 | |
| 3-Chlorobenzaldehyde | 213 | 18 | 255 | 228 | | 216 | |
| 2-Bromobenzaldehyde | 230 | 22 | 203 | 214 | | 240 | |
| 3-Bromobenzaldehyde | 234 | | 256 | 205 | | 220 | |
| 2-Aminobenzaldehyde | | 40 | 250 | 247 | | 219 | Oxime, 135 (p. 28) |
| 4-(Diethylamino)benzal-dehyde | | 41 | 206 | 214 | | | Oxime, 93 (p. 28) |
| 2-Nitrobenzaldehyde | | 44 | 250d | 256 | | 263 | |
| 4-Chlorobenzaldehyde | 214 | 47 | 268 | 231 | | 239 | |
| Trichloroethanal hydrate (Chloral hydrate) | | 57 | 131 | 90d | | | |
| 4-Bromobenzaldehyde | | 57 | 257* 128* | 228 | | 208 | *Polymorphs |
| 3-Iodobenzaldehyde | | 57 | 258 | 226 | | 213 | |
| 3-Nitrobenzaldehyde | | 58 | 293d | 246 | 198 | 247 | |
| 2,4-Dichlorobenzal-dehyde | | 71 | 226 | | | 256 | Oxime, 136 (p. 28) |
| 4-(Dimethylamino)benzal-dehyde | | 74 | 237 | 222 | | 182 | |
| 4-Iodobenzaldehyde | | 78 | 257 | 224 | | 201 | |
| 4-Nitrobenzaldehyde | | 106 | 320 | 220 | 190 | 249 | |
| 5-Bromo-2-hydroxy-benzaldehyde (5-Bromosalicylaldehyde) | | 106 | 292 | 297d | | | Oxime, 126 (p. 28). See Table 31 |

**Table 6.** Amides (primary), imides, ureas, thioureas and guanidines

| | M.p. | Xanthyl deriv. (p. 28) | Carboxylic acid (p. 28) | Notes |
|---|---|---|---|---|
| Methanamide (Formamide) | 3 | 184 | | B.p. 193 |
| Ethyl carbamate (Urethane) | 49 | 169 | | |
| Methyl carbamate | 54 | 193 | | |
| Propynamide (Propiolic amide) | 61 | | | Unsaturated |
| Propanamide | 81 | 214 | | |
| Ethanamide (Acetamide) | 82 | 240 | | |
| Propenamide (Acrylamide) | 84 | | | Unsaturated |
| 2-Phenylpropanamide | 92 | 158 | | |
| Maleimide | 93 | | 130 | Unsaturated |
| Heptanamide | 96 | 154 | | |

51

**Table 6:** Amides (primary), imides, ureas, thioureas and guanidines (*cont.*)

| | M.p. | Xanthyl deriv. (p. 28) | Carboxylic acid (p. 28) | Notes |
|---|---|---|---|---|
| Semicarbazide | 96 | | | Ethanal semicarbazone, 163 |
| Hexanamide | 100 | 160 | | |
| N-Methylurea | 101 | 230 | | Ethanoyl deriv., 180 |
| 3-Phenylpropanamide | 105 | 189 | 48 | |
| Hexadecanamide | 106 | 141 | 62 | |
| Pentanamide | 106 | 167 | | |
| Octanamide | 106 | 148 | 16 | |
| Octadecanamide | 108 | 141 | 70 | |
| Butanamide | 116 | 186 | | |
| Chloroethanamide | 120 | 209 | * | *Hydrolysis gives hydroxyethanoic acid, 80 |
| Cyanoethanamide | 123 | 223 | | |
| Succinimide | 125 | 246 | 185 | |
| Benzamide | 128 | 223 | 122 | |
| 2-Methylpropanamide | 128 | 211 | | |
| Urea | 132 | 274 | | Nitrate, 163 |
| 3-Methylbutanamide | 135 | 183 | | |
| 2-Hydroxybenzamide (Salicylamide) | 139 | | 158 | See Table 31 |
| 2-Furamide | 142 | 209 | 133 | |
| 2-Toluamide | 142 | 200 | 105 | |
| N-Phenylurea | 147 | 225 | | |
| N-Phenylthiourea | 154 | | | |
| Benzilamide | 155 | | 150 | |
| Phenylethanamide | 157 | 196 | 76 | |
| 4-Toluamide | 160 | 225 | 180 | |
| Malonamide | 170 | 270 | 133d | |
| Guanidine hydrochloride | 172 | | | Cold conc. $HNO_3$ + $H_2SO_4 \rightarrow$ nitroguanidine, 230d |
| 4-Ethoxyphenylurea (Dulcin) | 173 | | | Heating above m.p. $\rightarrow$ di-(4-ethoxy-phenyl)urea, 235 |
| Thiourea | 180 | | | Heating at m.p. $\rightarrow NH_4$ CNS which with aq. $FeCl_3$ gives red colour; N,S-diethanoyl deriv. 153; N-benzoyl deriv. 171 |
| NN-Dimethylurea | 182 | 250 | | |
| Thiosemicarbazide | 182 | | | Ethanoyl deriv., 165; benzaldehyde thio-semicarbazone, 160 |
| 1,1-Diphenylurea (Carbanilide) | 189 | 180 | | |
| 2-Tolylurea | 191 | 228 | | |
| Biuret* | 192d | 260 | | *$H_2$ N·CO·NH·CO·NH$_2$. With trace of $CuSO_4$ + dil. NaOH $\rightarrow$ red colour; an excess of $CuSO_4$ gives violet colour |
| Guanidine carbonate | 197 | | | Cold conc. $HNO_3$ + $H_2SO_4 \rightarrow$ nitroguanidine, 230d |
| 4-Nitrobenzamide | 201 | 232 | 240 | |
| 4-Nitrophthalimide | 202 | | 165 | |
| N-Cyanoguanidine (Dicyandiamide) | 208 | | | |
| Guanidine nitrate | 214 | | | Cold conc. $HNO_3$ + $H_2SO_4 \rightarrow$ nitroguanidine, 230d |
| 3-Nitrophthalimide | 218 | | 218 | |
| Phthalamide | 220 | | 200d | Loses $NH_3$ near its m.p. to give phthalimide, 233 |
| 2-Benzoic sulphimide (Saccharin) | 230 | 199 | | |
| Phthalimide | 233 | 177 | 200d | |
| Succinamide | 260d | 275 | 125 | |

**Table 7.** Amides, *N*-substituted. *N*-Substituted amides listed as ethanoyl and benzoyl derivatives of amines and amino-acids in Tables 8—11 and 13 should be consulted together with the following list. Hydrolysis (p. 28) should be followed by identification of the acid and amine.

| | B.p | | M.p. |
|---|---|---|---|
| *NN*-Dimethylmethanamide | | *N*-(4-Methoxyphenyl)ethanamide | 127 |
| (*NN*-Dimethylformamide) | 153 | *N*-(2,4-Dimethylphenyl)ethanamide | 130 |
| *NN*-Dimethylethanamide | | *N*-Phenylbromoethanamide | 131 |
| (*NN*-Dimethylacetamide) | 165 | *N*-(4-Ethoxyphenyl)ethanamide | |
| *NN*-Diethylmethanamide | | (Phenacetin) | 135[c] |
| (*NN*-Diethylformamide) | 176 | *N*-2-Tolyl benzamide | 142 |
| | | *N*-Phenyl-4-toluamide | 144 |
| | | *N*-Chlorobutanedioic imide | 150 |
| | M.p. | *N*-4-Tolylethanamide | 153[d] |
| *N*-Phenylmethanamide (Formanilide) | 46 | *N*-Methyl-*N*-(4-nitrophenyl)ethanamide | 153 |
| 3-Oxobutanamide (Acetoacetamide) | 54 | *N*-4-Tolylbenzamide | 158 |
| *N*-Ethyl-*N*-phenylethanamide | | *N*-1-Naphthylethanamide | 160 |
| (*N*-Ethylacetanilide) | 54[a] | *N*-1-Naphthylbenzamide | 161 |
| *N*-Phenyloctanamide | 57 | *N*-Phenylbenzamide (Benzanilide) | 163[e] |
| *N*-Phenylpentanamide | 63 | *N*-(4-Bromophenyl)ethanamide | |
| *N*-Phenyl-3-oxobutanamide | | (4-Bromoacetanilide) | 167 |
| (Acetoacetanilide) | 85 | *N*-Bromobutanedioic imide | |
| *N*-(2-Methoxyphenyl)ethanamide | 87 | (*N*-Bromosuccinimide) | 174 |
| *N*-Phenyloctadecanamide | 94 | *N*-(4-Chlorophenyl)ethanamide | 178 |
| *NN*-Diphenylethanamide | 101 | *NN'*-Diethanoyl-1,2-diaminobenzene | 186 |
| *N*-Methyl-*N*-phenylethanamide | 102 | *N*-Benzoylglycine (Hippuric acid) | 187 |
| *N*-Phenylpropanamide | 106[b] | *NN'*-Diethanoyl-1,3-diaminobenzene | 190 |
| *N*-Phenylethanamide (Acetanilide) | 114 | *N*-Phenyl-4-chlorobenzamide | 194 |
| *N*-Phenyl-2-chlorobenzamide | 118 | *N*-(4-Hydroxyphenyl)-*N*-methylethanamide | 240 |
| *N*-Phenyl-2-toluamide | 125 | Barbituric acid | 245 |
| *N*-Phenyl-3-toluamide | 126 | *NN'*-Diethanoyl-1,4-diaminobenzene | 303 |

[a] Conc. $HNO_3$ + $H_2SO_4$ at 40° → 4-nitro deriv., 118.
[b] Conc. $HNO_3$ + $H_2SO_4$ at 0° → 4-nitro deriv., 182.
[c] Warm 10% $HNO_3$ → 3-nitro deriv., 103.
[d] $Br_2$-ethanoic acid → 3-bromo deriv., 117.
[e] $Br_2$-ethanoic acid → 4-bromo deriv., 202.

**Table 8.** Amines, primary aliphatic

| | B.p | M.p | Ethanoyl deriv. (p. 29) | Benz-oyl deriv. (p. 29) | Tolu-ene-4-sulph-onyl deriv. (p. 29) | 2,4-Di-nitro-phenyl deriv. (p. 30) | Notes |
|---|---|---|---|---|---|---|---|
| Methylamine | −7 | | 28 | 80 | 77 | 178 ⎫ | Normally supplied as an |
| Ethylamine | 17 | | | 69 | 62 | 113 ⎬ | aqueous solution |
| 2-Aminopropane | 32 | | | 100 | 51 | 95 ⎭ | |
| 1,1-Dimethylethylamine | 46 | | | 134 | | | Picrate, 198 (p. 30) |
| (t-Butylamine) | | | | | | | |
| Propylamine | 49 | | | 85 | 52 | 97 | |
| Prop-2-enylamine | 58 | | | Oil | 64 | 76 | Picrate, 140 (p. 30) |
| (Allylamine) | | | | | | | |
| 1-Methylpropylamine | 63 | | | 76 | 55 | | Picrate, 140 (p. 30) |
| (s-Butylamine) | | | | | | | |
| 2-Methylpropylamine | 69 | | | 57 | 78 | 94 | |
| (Isobutylamine) | | | | | | (80) | |

53

**Table 8.** Amines, primary aliphatic (*cont.*)

| | B.p | M.p | Ethanoyl deriv. (p. 29) | Benz- oyl deriv. (p. 29) | Tolu- ene-4- sulph- onyl deriv. (p. 29) | 2,4-Di- nitro- phenyl deriv. (p. 30) | Notes |
|---|---|---|---|---|---|---|---|
| Butylamine | 77 | | | 42 | 65 | 90 | Picrate, 151 (p. 30) |
| 3-Methylbutylamine (Isopentylamine) | 96 | | | | 65 | 91 | Picrate, 138 (p. 30) |
| Pentylamine | 104 | | | | | 81 | Picrate, 139 (p. 30) |
| Ethylenediamine | 116 | 8 | 172 | 249 | 160 | 306 | |
| Propane-1,2-diamine (Propylenediamine) | 119 | | 139 | 192 | 103 | | |
| Hexylamine | 130 | | | 40 | | 39 | Picrate, 127; benzene- sulphonamide, 96 (p. 29) |
| Cyclohexylamine | 134 | | 104 | 147 | 87 | 156 | |
| Propane-1,3-diamine (Trimethylenediamine) | 136 | | 126 | 147 | 148 | | |
| Butane-1,4-diamine (Tetramethylenediamine) | 159 | 27 | 137 | 177 | 224 | | |
| 1-Aminopropan-2-ol (Isopropanolamine) | 163 | | | | | | Picrate, 142; see Table 3 |
| 2-Aminoethanol | 171 | | | | Oil | | Picrate, 159; see Table 3 |
| Benzylamine | 184 | | 60 | 106 | 116 | 116 | |
| 1-Phenylethylamine | 185 | | 57 | 120 | | 118 | |
| 3-Aminopropanol | 188 | | | | | | Picrate, 222; see Table 3 |
| 2-Phenylethylamine | 197 | | 51 | 116 | 64 | 154 | |
| (−)-Menthylamine | 205 | | 145 | 156 | | | Picrate, 215 (p. 30) |
| Dodecylamine | 247 | 27 | | | 73 | | |
| Diphenylmethylamine (Benzhydrylamine) | 303 | | 146 | 172 | | | |
| Hexane-1,6-diamine (Hexamethylenediamine) | 204 | 42 | | 155 | | | Picrate, 220 (p. 30) |

**Table 9.** Amines, primary aromatic (C, H, (O) and N)

| | B.p | M.p | Etha- noyl deriv. (p. 29) | Benz- oyl deriv. (p. 29) | Tolu- ene-4- sulph- onyl deriv. (p. 29) | 2,4-Di- nitro- phenyl deriv. (p. 30) | Notes |
|---|---|---|---|---|---|---|---|
| Aniline | 184 | | 114 | 163 | 103 | 156 | |
| 2-Toluidine | 200 | | 109 | 144 | 110 | 120 | |
| 3-Toluidine | 203 | | 66 | 125 | 114 | 159 | |
| 2-Ethylaniline | 210 | | 112 | 147 | | | Picrate, 194 (p. 30) |
| 2,4-Dimethylaniline | 212 | | 130 | 192 | 181 | 156 | |
| 2,5-Dimethylaniline | 213 | 15 | 139 | 140 | 119 (232) | 150 | |
| 2,6-Dimethylaniline | 215 | | 177 | 168 | 212 | | |
| 4-Ethylaniline | 216 | | 94 | 151 | 104 | | |
| 2-Methoxyaniline (*o*-Anisidine) | 218 | 5 | 85 | 60 | 127 | 151 | |

**Table 9.** Amides, primary aromatic (C, H, (O) and N) (*cont.*)

| | B.p | M.p | Etha-noyl deriv. (p. 29) | Benz-oyl deriv. (p. 29) | Tolu-ene-4-sulph-onyl deriv. (p. 29) | 2,4-Di-nitro-phenyl deriv. (p. 30) | Notes |
|---|---|---|---|---|---|---|---|
| 3,5-Dimethylaniline | 220 | | 144 | 136 | | | |
| 2,3-Dimethylaniline | 221 | | 134 | 189 | | | |
| 2-Ethoxyaniline (*o*-Phenetidine) | 228 | | 79 | 104 | 164 | 164 | |
| 2,4,6-Trimethylaniline | 229 | | 216 | 204 | 167 | | |
| 3-Ethoxyaniline (*m*-Phenetidine) | 248 | | 96 | 103 | 157 | | |
| 2-Aminoacetophenone | 250d | 20 | 76 | 98 | 148 | | See Table 27 |
| 3-Methoxyaniline (*m*-Anisidine) | 251 | | 80 | | 68 | 138 | |
| 4-Ethoxyaniline (*p*-Phenetidine) | 254 | 2 | 135 | 173 | 106 | 118 | |
| Methyl 2-aminobenzoate (Methyl anthranilate) | 255d | 25 | 101 | 100 | | | See Table 19 |
| 4-(Diethylamino)aniline | 262 | | 104 | 172 | | | |
| Ethyl 2-aminobenzoate (Ethyl anthranilate) | 265d | 13 | 61 | 98 | 112 | 164 | See Table 19 |
| 4-Toluidine | 201 | 45 | 153 (147) | 158 | 118 | 136 | |
| 3,4-Dimethylaniline | | 48 | 99 | 118 | 154 | 141 | |
| 2-Phenylaniline | | 49 | 118 | 102 | | | |
| 1-Naphthylamine | | 50 | 160 | 161 | 157 | 190 | Carcinogenic |
| 4-Phenylaniline | | 53 | 175 | 233 | 255 | | |
| 4-(Dimethylamino)aniline | | 53 | 131 | 228 | | | |
| 4-Methoxyaniline (*p*-Anisidine) | | 57 | 127 | 154 | 114 | 141 | |
| 2-Aminopyridine | | 58 | 71 | 169 | | | |
| 1,3-Phenylenediamine | | 63 | 190 di 88* mono | 240 di 125* mono | 172 | 172 | *More easily prepared than the di-amides |
| 3-Aminopyridine | | 64 | 133 | 119 | | | |
| 2-Nitroaniline | | 71 | 93 | 98 | 115 (142) | | Orange |
| 4-Aminodiphenylamine | | 75 | 158 | 203 | | 190 | |
| 2,4-Dimethyl-6-nitroaniline | | 76 | 176 | 185 | | | Orange |
| 4-Methyl-3-nitroaniline | | 78 | 145 | 172 | 164 | | Yellow |
| 2,4-Diaminophenol | | 79 | 180 tri 220 di-*N* | 253 di-*N* | | | |
| 2,5-Dimethoxyaniline | | 83 | 91 | 85 | 80 | 186 | |
| 2,5-Diethoxyaniline | | 86 | 116 | 88 | 132 | | |
| 4-Methyl-1,2-phenylenediamine (3,4-Diaminotoluene) | | 89 | 210 | 263 | 140 mono | | |
| Ethyl 4-aminobenzoate (Benzocaine) | | 90 | 110 | 148 | | | See Table 19 |
| 2-Methyl-3-nitroaniline | | 92 | 158 | 168 | | | Pale yellow |
| 2-Methyl-6-nitroaniline | | 97 | 158 | 167 | 122 | | Orange |
| 3-Aminoacetophenone | | 99 | 128 | | 130 | | See Table 27 |
| 4-Methyl-1,3-phenylenediamine (2,4-Diaminotoluene) | | 99 | 224 | 224 | 192 | 184 | |

**Table 9.** Amides, primary aromatic (C, H, (O) and N) (*cont.*)

| | B.p | M.p | Etha-noyl deriv. (p. 29) | Benz-oyl deriv. (p. 29) | Tolu-ene-4-sulph-onyl deriv. (p. 29) | 2,4-Di nitro-phenyl deriv. (p. 30) | Notes |
|---|---|---|---|---|---|---|---|
| 1,2-Phenylenediamine | | 102 | 186 | 301 | 202 | | |
| 4-Aminoacetophenone | | 106 | 167 | 205 | 203 | | |
| 2-Methyl-5-nitroaniline | | 107 | 151 | 183 | | | |
| 2-Naphthylamine | | 112 | 143 | 162 | 133 | 179 | Carcinogenic |
| 3-Nitroaniline | | 114 | 154 | 155 | 138 | 193 | Yellow |
| 4-Methyl-2-nitroaniline | | 117 | 96 | 148 | 166 (146) | | Red |
| 3-Aminophenol | | 122 | 101 di 148 mono-N | 153 di 174 mono-N | 157 mono | | See Table 31 Mono-O-benzoyl, 153 |
| 2,4-Dimethyl-5-nitroaniline | | 123 | 159 | 200 | 192 | | |
| 4-Aminobenzophenone | | 124 | 153 | 152 | | | |
| 4,4'-Diaminodiphenyl (Benzidine) | | 127 | 317 di 199 mono | 352 di 203 mono | 243 | | Carcinogenic |
| 4,4'-Diamino-3,3'-dimethyl-diphenyl (o-Tolidine | | 129 | 314 di 103 mono | 265 di 198 mono | | | Picrate, 185 (p. 30) Carcinogenic |
| 2-Methyl-4-nitroaniline | | 130 | 202 | 178 | 174 | | Light yellow |
| 1,4-Phenylenediamine | | 140 | 303 di 162 mono | 338 di 128 mono | 266 | 177 | |
| 2-Aminobenzoic acid (Anthranilic acid) | | 144 | 185 | 182 | 217 | | See Table 17 |
| 4-Nitroaniline | | 147 | 216 | 199 | 191 | 186 | Yellow |
| 3,5-Dinitro-2-hydroxyaniline (2-Amino-4,6-dinitrophenol, Picramic acid) | | 168 | 201 N- 193 O- | 230 N- 220 O- | 191 N- | | Red. See Table 31 |
| 2-Aminophenol | | 174d | 124 di 201* mono | 165 N- 185 O- | 146 (139) | | See Table 31 *Normal product of acetylation |
| 3-Aminobenzoic acid | | 174 | 250 | 248 | | | See Table 17 |
| 2,4-Dinitroaniline | | 180 | 120 | 202 | 219 | | Yellow; gives red colour with dil. NaOH and propanone(acetone) |
| 4-Aminophenol | | 184d | 150 di 168 N- | 234 di 216 N- | 168 di 252 N- 142 O- | | See Table 31 |
| 4-Aminobenzoic acid | | 186 | 252 | 278 | 223 | | See Table 17 |
| 2,4,6-Trinitroaniline | | 190 | 230 | 196 | | | Yellow |
| 4-Aminophenylethanoic acid | | 200 | 170 | 198 | | | See Table 17 |

**Table 10.** Amines, primary aromatic (C, H, (O), N and halogen or S)

| | B.p. | M.p. | Ethanoyl deriv. (p. 29) | Benzoyl deriv. (p. 29) | Toluene-4-sulphonyl deriv. (p. 29) | 2,4-Dinitrophenyl deriv. (p. 30) | Notes |
|---|---|---|---|---|---|---|---|
| 4-Fluoroaniline | 188 | | 152 | 185 | | | |
| 2-Chloroaniline | 208 | | 88 | 99 | 105 | 150 | |
| 2-Chloro-4-methylaniline | 223 | | 118 | 138 | 103 | | |
| 3-Chloroaniline | 230 | | 73 | 122 | 138 | 184 | |
| 2-Aminothiophenol | 234 | 26 | 135 | 154 | | | |
| 2-Bromo-4-methylaniline | 240 | 26 | 118 | 149 | | | |
| 4-Chloro-2-methylaniline | 241 | 29 | 140 | 172 | 145 | | |
| 3-Chloro-2-methylaniline | 245 | | 159 | 173 | | | |
| 2-Bromoaniline | 250 (229) | 32 | 100 | 116 | 90 | 161 | |
| 3-Bromoaniline | 251 | 18 | 88 | 135 | | 178 | |
| 3-Bromo-4-methylaniline | 254 | 25 | 113 | 132 | | | |
| 3-Iodoaniline | 145/15 mm | 33 | 119 | 157 | 128 | | |
| 4-Aminothiophenol | 140/16 mm | 46 | 144 di 154 N- | 180 N- | | | |
| 2,5-Dichloroaniline | 251 | 50 | 133 | 120 | | | |
| 2,5-Dibromoaniline | | 51 | 172 | | | | |
| 4-Bromo-2-methylaniline | 240 | 59 | 157 | 115 | | | |
| 2-Iodoaniline | | 60 | 110 | 139 | | | |
| 2,4-Dichloroaniline | | 63 | 146 | 117 | 126 | 116 | |
| 4-Bromoaniline | | 66 | 167 | 202 | 101 | 158 | |
| 4-Iodoaniline | | 67 | 184 | 222 | | 176 | |
| 4-Chloroaniline | | 70 | 178 | 192 | 95 | 167 | |
| 2,4,6-Trichloroaniline | | 78 | 205 | 174 | | | Insol. in HCl; diazotize in ethanol and $H_2SO_4$ |
| 2,6-Dibromo-4-methylaniline | | 78 | 183* mono | | | | *Diethanoyl deriv., 101 |
| 2,4-Dibromoaniline | | 79 | 146 | 134 | 134 | | |
| 2,6-Dibromoaniline | | 83 | 210* mono | | | | *Diethanoyl deriv., 110 Pricrate, 123 (p. 30) |
| 4-Chloro-1,3-phenylenediamine (2,4-Diaminochlorobenzene) | | 88 | 243* di | 178 | 215 | | *Monoethanoyl deriv., 170 |
| 2-Bromo-4-nitroaniline | | 105 | 129 | 160 | | | Yellow |
| 2-Chloro-4-nitroaniline | | 108 | 139 | 161 | 164 | | Yellow |
| 4-Bromo-2-nitroaniline | | 111 | 104 | 137 | | | Orange |
| 4-Chloro-2-nitroaniline | | 116 | 104 | | 110 | | Orange |
| 2,4,6-Tribromoaniline | | 119 | 232 mono 127 di | 198 | | | Insol. in HCl; diazotize in ethanol and $H_2SO_4$ |
| 4-Aminobenzenesulphonamide (Sulphanilamide) | | 165 | 219 4-N- 254 di | 284 4-N- 268 di | | | See Table 33 |

## Table 11. Amines, secondary

| | B.p. | M.p. | Ethanoyl deriv. (p. 29) | Benzoyl deriv. (p. 29) | Toluene-4-sulphonyl deriv. (p. 29) | 2,4-Dinitrophenyl deriv. (p. 30) | Notes |
|---|---|---|---|---|---|---|---|
| Dimethylamine | 7 | | | 42 | 79 | 87 | Normally in aq. solution |
| Diethylamine | 55 | | | 42 | 60 | 80 | |
| Di-isopropylamine | 84 | | | | | | Picrate, 140 (p. 30) |
| Pyrrolidine | 89 | | | | 123 | | Picrate, 112 (p. 30) |
| Piperidine | 106 | | | 48 | 103 | 93 | Miscible with water |
| Dipropylamine | 110 | | | | | 40 | Picrate, 75 (p. 30) |
| 2-Methylpiperidine | 117 | | | 45 | 55 | | Picrate, 135 (p. 30) |
| 3-Methylpiperidine | 124 | | | | | | Picrate, 137 (p. 30) |
| Morpholine | 130 | | | 75 | 147 | | |
| Di-isobutylamine | 139 | | 86 | | | 112 | |
| N-Methylcyclohexyl-amine | 146 | | | 85 | | | Picrate, 170 (p. 30) |
| 2-Ethylpiperidine | 146 | | | | | | Picrate, 133 (p. 30) |
| Dibutylamine | 159 | | | | | | Picrate, 60 (p. 30) |
| N-Methylaniline | 194 | | 101 | 63 | 95 | 167 | Picrate, 145 |
| N-Ethylaniline | 205 | | 54 | 60 | 88 | 95 | |
| N-Methyl-3-toluidine | 206 | | 66 | | | | |
| N-Methyl-2-toluidine | 207 | | 56 | 66 | 120 | 155 | |
| N-Methyl-4-toluidine | 208 | | 83 | 53 | 60 | | |
| N-Ethyl-2-toluidine | 214 | | | 72 | 75 | 114 | |
| N-Ethyl-4-toluidine | 217 | | | 40 | 71 | 120 | |
| N-Ethyl-3-toluidine | 221 | | | 72 | | | Picrate, 132 (p. 30) |
| N-Propylaniline | 222 | | 48 | | 56 | | |
| 1,2,3,4-Tetrahydro isoquinoline | 233 | | 46 | 129 | | | Picrate, 200 (p. 30) |
| N-Butylaniline | 240 | | | 56 | 54 | | |
| 1,2,3,4-Tetrahydro-quinoline | 250 | 20 | | 75 | | | Picrate, 141 (p. 30) |
| Dicyclohexylamine | 254 | 20 | 103 | 153 | 119 | | |
| Di-(2-hydroxyethyl)amine (Diethanolamine) | 270 | 28 | | | 99 | | Picrate, 110 (p. 30) |
| N-Methyl-1-naphthyl-amine | 193 | | 95 | 121 | 164 | | |
| Dibenzylamine | 300 | | | 112 | 158 | 105 | |
| N-Benzylaniline | 306 | 37 | 58 | 107 | 148 | 168 | |
| Piperazine hexahydrate | | 44* | 134 di 52 mono | 191 di 75 mono | 173 mono | | *Anhydrous, 104, and is sol. in water but only slightly in ether |
| Indole | | 52 | | 68 | | | Picrate, 187 |
| Diphenylamine | | 54 | 101 | 180 (107*) | 142 | | *Resolidifies at 135 and melts at 180 |
| N-Phenyl-1-naphthylamine | | 62 | 115 | 152 | | | |
| N-Methyl-3-nitroaniline | | 66 | 95 | 105 (156) | | | |
| N-Phenyl-2-naphthylamine | | 108 | 93 | 148 (111) | | | |
| N-Methyl-4-nitroaniline | | 152 | 153 | 112 | | | |
| Carbazole | | 243 | 69 | 98 | 137 | | Very weakly basic |

**Table 12.** Amines, tertiary

| | B.p. | M.p. | Meth-iodide (p. 30) | Picrate (p. 30) | Notes |
|---|---|---|---|---|---|
| Trimethylamine | 3 | | 230 | 216 | |
| Triethylamine | 89 | | 280 | 173 | |
| Pyridine | 116 | | 117 | 167 | |
| 2-Methylpyridine (α-Picoline) | 129 | | 230 | 169 | |
| 2-(Dimethylamine)ethanol | 135 | | | 96 | |
| 2,6-Dimethylpyridine (2,6-Lutidine) | 143 | | 238 | 161 | |
| 3-Methylpyridine (β-Picoline) | 144 | | 92 | 150 | |
| 4-Methylpyridine (γ-Picoline) | 144 | | 152 | 167 | |
| 2-Ethylpyridine | 149 | | | 187 | |
| Tripropylamine | 156 | | 208 | 117 | |
| 2,4-Dimethylpyridine (2,4-Lutidine) | 158 | | 113 | 180 | |
| 2-(Diethylamino)ethanol | 161 | | 249d | 79 | |
| 4-Ethylpyridine | 164 | | | 168 | |
| 2-Chloropyridine | 166 | | | | Methyl toluene-4-sulphonate salt, 120 (p. 30) |
| 3,5-Dimethylpyridine (3,5-Lutidine) | 170 | | | 238 | |
| 2,4,6-Trimethylpyridine (2,4,6-Collidine) | 172 | | | 156 | |
| NN-Dimethyl-2-toluidine | 185 | | 210 | 122 | |
| NN-Dimethylaniline | 193 | | 218* | 162 | *Sublimes. 4-Nitroso deriv., 87 (p. 30) |
| N-Ethyl-N-methylaniline | 201 | | 125 | 134 | 4-Nitroso deriv., 66 (p. 30) |
| NN-Dimethyl-4-toluidine | 210 | | 215* | 130 | *Sublimes |
| Tributylamine | 211 | | 180 | 107 | |
| NN-Dimethyl-3-toluidine | 212 | | 177 | 131 | |
| NN-Diethylaniline | 216 | | 104 | 142 | 4-Nitroso deriv., 84 (p. 30) |
| NN-Diethyl-4-toluidine | 229 | | 184 | 110 | |
| NN-Diethyl-3-toluidine | 231 | | | 97 | |
| Quinoline | 238 | | 72* | 203 | *Hydrate; anhydrous, 133 |
| Isoquinoline | 243 | 24 | 159 | 222 | |
| NN-Dipropylaniline | 245 | | 156 | 261 | |
| 2-Methylquinoline (Quinaldine) | 247 | | 195 | 191 | |
| 8-Methylquinoline | 248 | | 193 | 203 | |
| 6-Methylquinoline | 258 | | 219 | 234 | |
| 4-Methylquinoline (Lepidine) | 262 | | 174 | 212 | |
| 2,4-Dimethylquinoline | 264 | | 252 | 193 | |
| N-Benzyl-N-methylaniline | 306 | | 164 | 127 | 4-Nitroso deriv., 44 (p. 30) |
| N-Benzyl-N-ethylaniline | 312d | | 161 | 111 | 4-Nitroso deriv., 62 |
| 4-(Diethylamino)benzaldehyde | | 41 | | | Yellow; see Table 5 |
| 4-Bromo-NN-dimethylaniline | | 55 | 185 | | |
| 2,6-Dimethylquinoline | | 60 | 244d | 191 | |
| NN-Dimethyl-3-nitroaniline | | 60 | 205 | 119 | Red |
| NN-Dibenzylaniline | | 71 | 135 | 131 | 4-Nitroso deriv., 91 (p. 30) |
| 4-(Dimethylamino)benzaldehyde | | 74 | | | See Table 5 |
| 4-(Dimethylamino)phenol | | 75 | | | See Table 31 |
| 8-Hydroxyquinoline | | 75 | 143 | 204 | See also Table 31 |
| 3-(Dimethylamino)phenol | | 85 | 182 | 162 | 4-Nitroso deriv., 169; see Table 31 |

**Table 12.** Amides, tertiary (*cont.*)

| | B.p. | M.p. | Meth-iodide (p. 30) | Picrate (p. 30) | Notes |
|---|---|---|---|---|---|
| Tribenzylamine | 92 | 184 | 190 | | |
| Acridine | 110 | 224 | 208 | | |
| 2,3-Dimethyl-1-phenylpyrazol-5-one (Phenazone, antipyrine) | 111 | | 181 | | 4-Nitroso deriv., 200 Absorbs bromine; gives orange colour with aq. FeCl$_3$ |
| Triphenylamine | 127 | | | | Not basic; nitration in ethanoic acid gives trinitro deriv., 280 |
| 4,4′-Di-(dimethylamino)benzophenone (Michler's ketone) | 174 | 105 | 156 | | Oxime, 233 (p. 39). See Table 27 |
| Hexamethylene tetramine | 280* | 190 | 179 | | *Sublimes |

**Table 13.** Amino-acids

| | Decomposition temp. | Benzoyl deriv. (p. 31) | 3,5-Dinitrobenzoyl deriv. (p. 31) | Toluene-4-sulphonyl deriv. (p. 31) | Notes |
|---|---|---|---|---|---|
| N-Phenylglycine | 126 | 63 | | | Ethanoyl deriv., 194 (p. 31) |
| (+)- or (−)-Ornithine | 140* | 188 di 240 mono | | | *Often a syrup. Picrate, 204 (p. 30) |
| 2-Aminobenzoic acid (Anthranilic acid) | 144 | 182 | 278 | | Ethanoyl deriv., 185 (p. 31); see Tables 9 and 17 |
| 3-Aminobenzoic acid | 174 | 248 | | | Ethanoyl deriv., 250 (p. 31); see Tables 9 and 17 |
| 4-Aminobenzoic acid | 186 | 278 | | | Ethanoyl deriv., 252 (p. 31); see Tables 9 and 17 |
| 3-Aminopropanoic acid (β-Alanine) | 196 | | 202 | 117 | |
| Glutamic acid | 199 | 157 | | 117 | Ethanoyl deriv., 185 (p. 31) |
| 4-Aminophenylacetic acid | 200 | 205 | | | Ethanoyl deriv., 170 (p. 31) |
| Proline | 203* | | 217 | | *Monohydrate, 190. Picrate, 135 |
| (+)- or (−)-Arginine | 207 | 298 mono 235 di | 150 | | Picrate, 206 (p. 30) |
| (+)- or (−)-Glutamic acid | 211 | 138 | 217 | 131 | |
| Sarcosine | 212 | 103 | 154 | 102 | |
| (+)- or (−)-Proline | 222 | 156 | | 133 | Picrate, 154 (p. 30) |
| (+)- or (−)-Lysine | 224 | 149 | 169 | | Picrate, 266 (p. 30) |
| (+)- or (−)-Asparagine | 226 | 189 | 196 | 175 | |
| Serine | 228 | 171 mono 124 di | 95 | 213 | |
| Glycine | 232 | 187 | 179 | 147 | Ethanoyl deriv., 206 (p. 31) |

**Table 13.** Amides-acids (*cont.*)

| | Decomposition temp. | Benzoyl deriv. (p. 31) | 3,5-Dinitrobenzoyl deriv. (p. 31) | Toluene-4-sulphonyl deriv. (p. 31) | Notes |
|---|---|---|---|---|---|
| Threonine | 235 | 174* di 176* mono | | | *Mixed m.p. 145 |
| Arginine | 238 | 230* | | | *Hydrate, 176. Picrate, 201 (mono), 196 (di) (p. 30) |
| (+)- or (−)-Cystine | 260 | 181 | 180 | 201 | |
| (+)- or (−)-Aspartic acid | 271 | 185 | | 140 | |
| Methionine | 281 (272) | 151 | | 105 | Ethanoyl deriv., 114 (p. 31) |
| Phenylalanine | 273 | 188 | 93 | 134 | |
| Tryptophan | 275 | 188 | 240 | 176 | |
| (+)- or (−)-Histidine | 277 | 249 | 189 | 203 | |
| 2-Amino-2-methylpropionic acid | 280* | 198 | | | *Sublimes |
| Aspartic acid | 280 | 165 | | | |
| (+)- or (−)-Methionine | 283 | 150 | 95 | | Ethanoyl deriv., 98 (p. 31) |
| (+)- or (−)-Tryptophan | 289 | 176 | 233d | 176 | |
| Isoleucine | 292 | 118 | | 140 | |
| Alanine | 295 | 166 | 177 | 139 | |
| (+)- or (−)-Alanine | 297 | 151 | | 133 | |
| Valine | 298 | 132 | 158 | 110 | |
| 2-Aminobutanoic acid | 307 | 147 | 194 | | |
| (+)- or (−)-Valine | 315 | 127 | 158 | 147 | |
| Tyrosine | 318 | 197 mono-*N* | 254 | 224 | |
| (+)- or (−)-Phenylalanine | 320 | 146 | 93 | 164 | |
| Norleucine | 327 | | | 124 | |
| Leucine | 332 | 141 | 187 | | Ethanoyl deriv., 157 |
| (+)- or (−)-Leucine | 337 | 107* | 187 | 124 | *Hydrate, 60 |
| (+)- or (−)-Tyrosine | 344 | 166 mono-*N* 211 di | | 188 mono-*N* 119 di | Ethanoyl deriv., 172 |
| Ornithine | | 267d mono 188 di | | 188 mono | |
| Lysine | | 249 mono 145 di | | | Picrate, 225d (mono) |

**Table 14.** Azo, azoxy, nitroso and hydrazine compounds

| | B.P. | Notes |
|---|---|---|
| *NN*-Dimethylhydrazine | 63 | Picrate, 146 (p. 30) |
| *NN'*-Dimethylhydrazine | 81 | Picrate, 148 (p. 30) |
| *N*-Methyl-*N*-phenylhydrazine | 227 | Benzoyl deriv., 153; ethanoyl deriv., 92 (p. 29); with benzaldehyde → hydrazone, 106 |
| *N*-Ethyl-*N*-phenylhydrazine | 237 | Hydrochloride, 146 |
| *N*-Ethyl-*N'*-phenylhydrazine | 238 | Benzoyl deriv., 100 (p. 26) |
| Phenylhydrazine (m.p. 19) | 243 | Benzoyl deriv., 168; with benzaldehyde →hydrazone, 158 |

| | M.p. | |
|---|---|---|
| *NN*-Diphenylhydrazine | 34 | Benzoyl deriv., 192; ethanoyl deriv., 184; with benzaldehyde → hydrazone, 122 |
| Azoxybenzene | 36 | Light yellow. Warming with conc. $H_2SO_4$ → 4-hydroxyazobenzene, 152 |
| 4-Tolylhydrazine | 66 | Benzoyl deriv., 146; ethanoyl deriv., 131; with benzaldehyde → hydrazone, 125 |
| Nitrosobenzene | 68 | Colourless solid but turns green on melting. With Sn + HCl → aniline (p. 40) |
| Azobenzene | 68 | Orange-red colour. Zn dust + ethanolic NaOH → hydrazobenzene, 131 |
| 4,4'-Azoxytoluene | 75 | Light yellow. Zn dust + ethanolic NaOH → 4,4'-azotoluene, 144 |
| *NN*-Diethyl-4-nitrosoaniline | 84 | Dark green, forms yellow hydrochloride; $KMnO_4$ → *NN*-diethyl-4-nitroaniline, 77 |
| *NN*-Dimethyl-4-nitrosoaniline | 85 | Dark green, forms yellow hydrochloride; $KMnO_4$ → *NN*-dimethyl-4-nitroaniline, 163 |
| 4-Nitrosophenol | 125d | Colourless Ethanoic-anhydride at 100° → ethanoyl deriv., 107 (yellow) |
| *NN'*-Diphenylhydrazine (Hydrazobenzene) | 127 | Dibenzoyl deriv., 162 (p. 26) |
| Benzeneazo-2-naphthol | 133 | Orange colour. See Table 31 |
| 4-Nitrosodiphenylamine | 145 | Green. Zn + ethanoic acid → *p*-amino deriv., 66 |
| 1,2-Di-(2-hydroxyphenyl)hydrazine (2,2'-Hydrazophenol) | 148 | Yellow. Benzoyl deriv., 186; see also Table 31 |
| 4-Hydroxyazobenzene | 152 | Yellow. Ethanoyl deriv., 84; benzoyl deriv., 138 (p. 41) |
| 2,4-Dinitrophenylhydrazine | 197d | Red. Ethanoyl deriv., 197; benzoyl deriv., 206 |
| 1-Nitroso-2-naphthol | 109 | Orange. Benzoyl deriv., 114. Cold dil.$HNO_3$ → I—$NO_2$ deriv., 103 |
| 4-Nitrophenylhydrazine | 157d | Orange. Picrate, 119; benzaldehyde → hydrazone, 192 |

**Table 15.** Carbohydrates

| | Approximate decomposition temperature | $[\alpha]_D$ (in water) Initial | Final | 4-N-Glycosylaminobenzoic acid (p. 32) | Ethaoate* (p. 31) | Osazone (p. 32) | Notes |
|---|---|---|---|---|---|---|---|
| Melibiose, monohydrate | 85 | +111 | +129 | | α147 β177 | 178 | Disaccharide |
| D-Ribose | 90 | −21 | −21 | 156 | | 164 | |
| D-Glucose, monohydrate | 90 | +112 | +52 | 134 | α112 β132 | 205d | Pentabenzoate, 179 (p. 26) |
| Maltose, monohydrate | 101 | +111 | +130 | | α125 β160 | 206 | Disaccharide |
| D-Fructose | 104 | −132 | −92 | | α 70 β109 | 205d | Pentabenzoate, 79 (p. 26) |
| L-Rhamnose, monohydrate | 105 | −8.6 | +8.2 | 170 | 99 | 182 | |
| L-Rhamnose, anhydrous | 123 | −8.6 | +8.2 | 170 | 99 | 182 | |
| D-Mannose | 132 | +29 | +14 | 182 | α 74 β115 | 205d | |
| D-Xylose | 144 | +93 | +18 | 181 | α 59 β126 | 160 | |
| D-Glucose, anhydrous | 146 | +112 | +52 | 134 | α112 β132 | 205d | |
| D- or L-Arabinose | 160 | −175 D-+190 L- | −105 D-+104 L- | 192 | α 94 β 86 | 166 | |
| L-Sorbose | 161 | −43 | −43 | | 97 | 162 | |
| Maltose, anhydrous | 165 | +111 | +130 | | α125 β160 | 206 | Dissaccharide |
| D-Galactose | 170 | +150 | +80 | 160 | α 95 β142 | 196 | |
| Sucrose | 185 | +66 | +66 | | 70 | 205 | Non-reducing disaccharide |
| Lactose, monohydrate | 203 | +90 | +55 | | α152 β 90 | 200d | Disaccharide |
| Cellobiose | 225 | +14 | +35 | | α229 β192 | 200 | Disaccharide |

*The preparation of the β-form is described on p. 31 but the melting points of both anomers are given because a small amount of the α-form is sometimes formed.

**Table 16.** Carboxylic acids (C, H and O), their acyl chlorides, anhydrides and nitriles

| | B.p. | M.p. | Chloride B.p. | Anhydride B.p. | Nitrile M.p. | Amide (p. 32) M.p. | Anilide (p. 32) M.p. | 4-Toluidide (p. 32) M.p. | 4-Bromophenacyl ester (p. 32) M.p. | 4-Phenylphenacyl ester (p. 32) M.p. | Notes |
|---|---|---|---|---|---|---|---|---|---|---|---|
| Methanoic (formic) | 100 | 8 | — | — | 26 | 3 | 50* | 53* | 140† | 74 | *By heating the acid with the amine †ArCO·CH$_2$·OH is often isolated; ester melts at 99 |
| Ethanoic (acetic) | 118 | 16 | 52 | 140 | 82 | 82 | 114 | 153 (147) | 85 | 111 | |
| Propanoic | 140 | | 80 | 168 | 97 | 81 | 106 | 126 | 61 | 102 | |
| Propenoic (Acrylic) | 140 | 13 | 75 | — | 78 | 84 | 104 | 141 | | | Unsaturated; polymerizes readily |
| Propynoic (Propiolic) | 144d | 18 | | | | 61 | 87 | | | | Unsaturated |
| 2-Methylpropanoic (Isobutyric) | 155 | | 92 | 182 | 108 | 128 | 105 | 108 | 77 | 89 | |
| 2-Methylpropenoic (Methacrylic) | 161 | 16 | 95 | | 90 | 102 | 87 | | | | Unsaturated |
| Butanoic (Butyric) | 163 | | 101 | 198 | 118 | 116 | 96 | 75 | 63 | 97 | |
| 2,2-Dimethylpropanoic (Pivalic) | 164 | 35 | 105 | 190 | 106 | 155 | 132 | 119 | 76 | 114 | |
| 2-Oxopropanoic (Pyruvic) | 165 | 13 | | | | 124 | 104 | 109 | | | See Table 26 |
| But-3-enoic (Vinylacetic) | 169 (163) | | 98 | | | 73 | 58 | | 60 | | Unsaturated |
| cis-But-2-enoic (Isocrotonic) | 169 | | | | | 102 | 101 | | | | Unsaturated |
| 2-Methylbutanoic | 177 | | 115 | 215 | 125 | 112 | 110 | 132 | 81 | 71 | |
| 3-Methylbutanoic (Isovaleric) | 177 | | 115 | 218 | 129 | 135 | 110 | 93 | 55 | 78 | |
| Pentanoic (Valeric) | 186 | | 126 | 229 | 140 | 106 | 63 | 107 | 68 | 64 | |
| 2-Ethylbutanoic | 195 | | 139 | | 145 | 112 | 127 | 74 | 75 | 77 | |
| 4-Methylpentanoic (Isocaproic) | 199 | | | | 155 | 120 | 112 | 116 | | 70 | |
| Hexanoic (Caproic) | 205 | | 153 | 254 | 163 | 100 | 94 | 63 | 77 | 68 | |
| Heptanoic | 223 | | 193 | 258 | 184 | 96 | 65 | 74 | 72 | 62 | |
| 2-Ethylhexanoic | 227 | | 88/20 mm | 150/8 mm | 75/9 mm | 102 | 89 | 81 | 72 | 53 | |
| Octanoic (Caprylic) | 237 | 16 | 195 | 281 | 205 | 106 | 57 | 70 | 66 | 67 | |
| 4-Oxopentanoic (Laevulinic, laevulic) | 245d | 33 | | | | 107 | 102 | 109 | 84 | | See Table 26 |
| Nonanoic | 254 | 12 | 215 | | 224 | 101 | 57 | 84 | 69 | 71 | |
| 2-Phenylpropanoic | 265 | | 97/12 mm | | 230 | 95 | | | | | |
| Decanoic (Capric) | 269 | 31 | 114/15 mm | 24* | 245 | 108 | 70 | 78 | 66 | | *M.p. |

Table of carboxylic acids and their derivative melting points (values in °C; * = M.p.; mm = b.p. at reduced pressure).

| Compound | m.p. | b.p. | (3) | (4) | (5) | (6) | (7) | (8) | (9) | Remarks |
|---|---|---|---|---|---|---|---|---|---|---|
| Undec-10-enoic (Undecylenic) | 24 | 275 | | 248 | 87 | 68 | 67 | 68 | 80 | Unsaturated |
| Undecanoic | 28 | 280 | 37* | 330d | 103 | 80 | 71 | 46 | 61 | *M.p. |
| cis-Octadec-9-enoic (Oleic) | 16 | 286d | 22* | 182 | 76 | 41 | 41 | 42 | | *M.p. Unsaturated |
| 2-Hydroxypropanoic (Lactic) | 18 | 122/15 mm | | | 78 | 58 | 107 | 113 | 145 | *M.p. |
| Dodecanoic (Lauric) | 44 | 145/18 mm | 42* | 280 | 100 | 78 | 87 | 76 | 84 | *M.p. |
| 3 Phenylpropanoic (Hydrocinnamic) | 48 | 225d | | 261 | 105 (82) | 96 | 135 | 104 | 95 | |
| trans-Octadec-9-enoic (Elaidic, trans-Oleic) | 51 | | 51* | | 93 | | | 65 | 73 | *M.p. Unsaturated |
| 4-Phenylbutanoic (4-Phenylbutyric) | 52 | 119/9 mm | | | 84 | | | 58 | 90 | |
| Tetradecanoic (Myristic) | 54 | 174/16 mm | 54* | 19* | 102 | 84 | 93 | 81 | 90 | *M.p. |
| Hexadecanoic (Palmitic) | 62 | 12* | 63* | 31* | 106 | 89 | 94 | 84 | 94 | *M.p. |
| 3-Methylbut-2-enoic (ββ-Dimethylacrylic) | 68 | 145 | | 140 | 107 | 98 | 102 | 104 | 145 | Unsaturated |
| Octadecanoic (Stearic) | 70 | 22* | 70* | 43* | 108 | 94 | 90 | 96 | 97 | *M.p. |
| But-2-enoic (Crotonic) | 72 | 126 | 246 | 118 | 158 | 118 | 132 | 96 | | Unsaturated |
| Phenylethanoic (Phenylacetic) | 76 | 210 | 72* | 232 | 157 | 118 | 136 | 89 | 63d | *M.p. |
| 2-Hydroxy-2-methylpropanoic (α-Hydroxybutyric) | 79 | | | | 98 | 136 | 133 | | | |
| Hydroxyethanoic (Glycollic) | 80 | 95/18 mm | 128* | 183 | 120 | 97 di 153 | 143 / 170 mono | 138 | 109 | Is often a syrup. *M.p. |
| Methylmaleic (Citraconic) | 92d | 214 | 214 | | 185 | 176 di / 153 mono | 170 mono | | | |
| Pentanedioic (Glutaric) | 98 | 218 | 56* | 286 | 175 | 224 | 218 | 137 | 152 | *M.p. |
| Phenoxyethanoic | 99 | 225 | 67* | 239 | 101 | 99 | 189 | 149 | 146 | *M.p. |
| 2-Carboxy-2-hydroxypentanedioic, monohydrate (Citric, monohydrate) | 100 | 64 | | | 210 | 199 | 189 | 148 | | Heat at 130° → anhyd. acid, 153 |
| Ethanedioic, dihydrate (Oxalic, dihydrate) | 100 | | | | 419 di / 219 mono | 246 di / 148 mono | 268 di / 169 mono | 242 | 166 | Heat → anhydrous acid, 189 |
| (−)-Hydroxysuccinic (Malic) | 100 | | | | 156 | 197 | 207 | 179 | 106 | |
| 2-Methoxybenzoic (o-Anisic) | 100 | 254 | 24* | | 129 | 78 | | 113 | 131 | *M.p. |
| Heptanedioic (Pimelic) | 104 | | | | 175 | 156 di / 108 mono | 206 | 137 | 146d | |
| 2-Toluic | 105 | 212 | 39* | 205 | 142 | 125 | 144 | 57 | 95 | *M.p. |

**Table 16.** Carboxylic acids (C, H and O), their acyl chlorides, anhydrides and nitriles (*cont.*)

| | M.p. | Chloride B.p. | Anhydride B.p. | Nitrile B.p. | Amide (p. 32) M.p. | Anilide (p. 32) M.p. | 4-Toluidide (p. 32) M.p. | 4-Bromophenacyl ester (p. 32) M.p. | 4-Phenylphenacyl ester (p. 32) M.p. | Notes |
|---|---|---|---|---|---|---|---|---|---|---|
| Nonanedioic (Azelaic) | 106 | 166/18 mm | | | 175 di / 94 mono | 187 di / 107 mono | 200 | 131 | 141 | |
| 3-Toluic | 111 | 218 | 71* | 212 | 96 | 126 | 118 | 108 | 137 | *M.p. |
| 3-Benzoylpropanoic | 116 | | | | 125 | 150 | | 98 | | See Table 26 |
| 3-Hydroxy-2-phenylpropanoic (Tropic) | 117 | | | | 169 | | | | | Ethanoate, 88 (p. 26) |
| 2-Hydroxy-2-phenylethanoic (Mandelic) | 118 | | | 21* | 133 | 151 | 172 | 113 | | *M.p. |
| Benzoic | 122 | 197 | 42* | 191 | 128 | 163 | 158 | 119 | 167 | *M.p. |
| 2-Benzoylbenzoic | 128* | 70† | | 83† | 165 | 195 | | | | *Monohydrate, 91. See Table 26. †M.p. |
| cis-Butenedioic (Maleic) | 130 (139) | 60* | 56* | 31* | 266 di / 181 mono | 187 | 142 | 168 | 168 | *M.p. Unsaturated |
| Propanedioic (Malonic) | 133d | | | 219 | 170 di / 50 mono | 225 | 252 | | 175 | |
| Decanedioic (Sebacic) | 133 | 182/16 mm | | | 210 di / 170 mono | 200 di / 122 mono | 201 | 147 | 140 | |
| 3-Phenylpropenoic (Cinnamic) | 133 | 35* | 136* | 255 | 147 | 151 | 168 | 146 | 183 | Unsaturated |
| Furoic | 133 | 173 | 73* | 20* / 146 | 142 | 124 | 108 | 139 | 91 | *M.p. |
| 1-Naphthylethanoic (1-Naphthylacetic) | 133 | 188/23 mm | | 183 | 181 | 155 | | 112 | | *M.p. |
| Hexa-2,4-dienoic (Sorbic) | 134 | 78/15 mm | | | 168 | 153 | | 129 | 141 | Unsaturated |
| 2-(Ethanoyloxy)benzoic (Acetylsalicylic, aspirin) | 135 | | 43* | | 138 | 136 | | | | *M.p. |
| 3-Phenylpropynoic (Phenylpropiolic) | 137 | 116/17 mm | | 41* | 100 | 126 | 142 | | | Unsaturated. *M.p. |

| Acid | M.p. | | | | | | | | | Remarks |
|---|---|---|---|---|---|---|---|---|---|---|
| meso-Dihydroxybutanedioic (meso-Tartaric) | 140 | | | 131* | 190 | 193 / mono | | | | *M.p. |
| Octanedioic (Suberic) | 142 | 162/15 | 65* | | 216 / di 125 / mono | 187 / di 128 / mono | 219 | 144 | 151 | *M.p. |
| 3,4,5-Trimethoxybenzoic | 144 | | | | 184 | 154 | 172 | 129 | 111 | |
| Diphenylethanoic | 148 | 56* | 98* | | 168 | 180 | 189 | 112 | 122 | *M.p. |
| Benzilic (Diphenylglycollic) | 150 | 193/27 mm | | 72* | 154 | 175 | 189 | 152 | | |
| 2-Carboxy-2-hydroxypentanedioic, anhydrous (Citric) | 153 | | | | 210* | 199* | 189* | 148 | 146 | *From amine and ester on prolonged heating |
| Hexanedioic (Adipic) | 153 | 130/18 | | 295 | 220 / di 126 / mono | 238 / di 151 / mono | 241 | 155 | 148 | |
| 2-Hydroxy-5-methylbenzoic (5-Methylsalicylic) | 153 | | | | 177 | | | 142 | | See Table 30 |
| 2-Hydroxybenzoic (Salicylic) | 158 | 20* | | 98* | 139 | 135 | 156 | 140 | 148 | See Table 30.  *M.p. |
| 1-Naphthoic | 161 | | 145* | 35* | 202 | 163 | | 135 | 126 | *M.p. |
| 2-Hydroxy-3-methylbenzoic (3-Methylsalicylic) | 163 | | | | 112 | | 164 | | | See Table 30 |
| Methylenebutanedioic (Itaconic) | 165 | | | | 192 | 190* | 117 | | | *By heating acid with an excess of amine  Unsaturated |
| Phenylbutanedioic | 167 | | | | 211 / 195 / di 171 / mono / 294d | 222 / 264 / di 180 / mono | 216 | | | |
| (+)-Dihydroxybutanedioic ((+)-Tartaric) | 169 | | | | | | | | 204d | Dibenzoate, 90 |
| Butynedioic (Acetylenedicarboxylic) | 179 | | | | 160 | 144 | | | | Unsaturated |
| 4-Toluic | 180 | 214 | 95* | 218 | 164 | 154 | 160 | 153 | 165 | *M.p. |
| 3,4-Dimethoxybenzoic (Veratric) | 181 | | | | 163 | 169 | | 124 | | |
| 4-Methoxybenzoic (p-Anisic) | 184 | 22* | 99* | 61* | 228 | 232 | 186 | 152 | 160 | *M.p. |
| 2-Carboxyphenylethanoic (Homophthalic) | 185 | 43* | 141* | | 192 | 171 | 191 | | | *M.p. |
| 2-Naphthoic | 185 | | 133* | 66* | 260 / di 157 / mono | 230 / di 148 / mono | 255 / di 179 / mono | 211 | 183 | *M.p. |
| Butanedioic (Succinic) | 185 | 20* | 119* | 54* | 192 / di 177 / mono | 226 / di 204 / mono | mono | 211 | 208 | *M.p. |
| (+) Camphoric | 187 | | | | | | | | | |

**Table 16.** Carboxylic acids (C, H and O), their acyl chlorides, anhydrides and nitriles (*cont.*)

| | M.p. | Chloride B.p. | Anhydride B.p. | Nitrile B.p. | Amide (p. 32) M.p. | Anilide (p. 32) M.p. | 4-Toluidide (p. 32) M.p. | 4-Bromophenacyl ester (p. 32) M.p. | 4-Phenylphenacyl ester (p. 32) M.p. | Notes |
|---|---|---|---|---|---|---|---|---|---|---|
| Ethanedioic, anhydrous (Oxalic) | 188 | 64 | | | 419 di / 219 mono | 246 di / 148 mono | 268 di / 169 mono | 242 | 166 | *M.p. Ethanoate, 158. See Table 30 |
| 1-Hydroxy-2-naphthoic | 195 | 85* | | | 202 | 154 | | | | *M.p. See Table 30 |
| 3-Hydroxybenzoic | 200 | | | 82* | 170 | 156 | 163 | 176 (168) | | *M.p. See Table 30 |
| 3,4-Dihydroxybenzoic (Protocatechnic) | 200d | | | 156* | 212 | 166 | | | | *M.p. See Table 30 |
| 2,5-Dihydroxybenzoic (Gentisic) | 200 | 98* | | | 218 | | | | | *M.p. See Table 30 |
| Benzene-1,2-dicarboxylic (Phthalic) | 200d | 275 | 131* | 141* | 220† | 254 di / 170 mono | 201 di / 155 mono | 153 | 167 | Loses water near m.p. to form anhydride. *M.p. †Loses NH₃ near m.p. to form imide, 233 |
| Dihydroxybutanedioic (Tartaric) | 205 | | | | 226 | 235 | | | | Dibenzoate, 112 |
| 4-Hydroxybenzoic | 213 | | | 113* | 162† | 196† | 204† | 191 | 240 | *M.p. †Difficult to prepare. See Table 30 |
| 2,4-Dihydroxybenzoic (β-Resorcylic) | 213d | | | | 222 | 126 | | | | *M.p. See Table 30 |
| Galactaric (Mucic) | 214d | | | 175* | 220 di / 192 mono | | | 225 | 149d | Tetra-ethanoate, 266 (p. 31) |
| 3-Hydroxy-2-naphthoic | 222 | 95* | | 188* | 217 | 243 | 221 | | | *M.p. Ethanoate, 184 See Table 30 |
| 3,5-Dihydroxybenzoic (α-Resorcylic) | 233 | | | | | | | | | See Table 30 |
| 3,4,5-Trihydroxybenzoic (Gallic) | 240d | | | | 245* (189) | 207* | | 134 | 198d | *Difficult to prepare. See Table 30 |
| *trans*-Butenedioic (Fumaric) | 287* | 160 | | 96† | 267 | 314 | | 256d | | *In a sealed tube; sublimes at 200. Unsaturated †M.p. |
| Benzene-1,4-dicarboxylic (Terephthalic) | >300* | 83† | | 222† | >350d | 334 | | 225 | 280 | *Sublimes †M.p. |
| Benzene-1,3-dicarboxylic (Isophthalic) | 345 | 41* | | 162* | 280 | 250 | | 179 | | *M.p. |

## Table 17. Carboxylic acids (C, H, O and halogen, N or S)

| | B.p. | M.p. | Amide (p. 32) | Anilide (p. 32) | 4-Toluidide (p. 32) | 4-Bromo-phenacyl ester (p. 32) | 4-Phenyl phenacyl ester (p. 32) | Notes |
|---|---|---|---|---|---|---|---|---|
| Trifluoroethanoic (Trifluoroacetic) | 72 | | 74 | 91 | | | | Acid chloride b.p. −27 |
| Thioacetic | 93 | 31 | 108 | 76 | 130 | | | Pale yellow; unpleasant odour |
| Fluoroethanoic | 167 | | 108 | 92 | 124 | | | |
| 2-Chloropropanoic | 186 | | 80 | 125 | 153 | | | |
| Dichloroethanoic | 194 | 5 | 98 | 99 | 125 | 99 | | |
| 2-Bromopropanoic | 203 | 25 | 123 | 98 | 92 | | | |
| 2-Bromobutanoic (2-Bromobutyric) | 217d | | 112 | | | | | |
| Mercaptoethanoic (Thioglycollic) | 123/29 mm | | 52 | 111 | 125 | | | |
| 3-Chloropropanoic | | 40 | 101 | 116 | 124 | | | |
| 2-Bromo-3-methylbutanoic | | 44 | 133 | 83 | 93 | | | |
| 2-Bromo-2-methylpropanoic | | 48 | 148 | 131 | 91 | | | |
| Bromoethanoic | 208 | 50 | 91 | 94 | 113 | | | |
| Trichloroethanoic | 196 | 57 | 141 | 137 | 162 | 105 | 116 | |
| Chloroethanoic | | 63 | 120 | | | | | |
| Cyanoethanoic | | 66 | 123 | 198 | | | | |
| Iodoethanoic | | 83 | 95 | 143 | 165 | | | |
| 2-Hydroxy-3-nitrobenzoic (3-Nitrosalicylic) hydrate | | 125 | 145 | | | | | |
| Pyridine-2-carboxylic (2-Picolinic) | | 138 | 107 | 76 | 104 | | | |
| 2-Chloro-4-nitrobenzoic | | 139 | 172 | 168 | | | | |
| 3-Nitrobenzoic | | 140 | 142 | 154 | 162 | 137 | 153 | |
| 2-Nitrophenylethanoic | | 141 | 161 | 131 | | | | |
| 4-Chloro-3-nitrobenzoic | | 142 | 172 | | | | | |
| 2-Chlorobenzoic | | 142 | 142 | 118 | 131 | 106 | 123 | |
| 2-Hydroxy-3-nitrobenzoic, anhyd. | | 144 | 145 | | | | | |
| 2-Aminobenzoic (Anthranilic) | | 144 | 109 | 131 | 151 | 172* | | *Two moles of reagent are required to form this derivative See Tables 9 and 13 |
| 2-Nitrobenzoic | | 146 | 175 | 155 | 203 | 107 | 140 | |
| 2-Bromobenzoic | | 150 | 155 | 141 | | 102 | 98 | |
| 4-Nitrophenylethanoic | | 152 | 198 | 198 | 210 | 207 | | |
| 3-Bromobenzoic | | 155 | 155 | 146 | | 126 | 155 | |

**Table 17.** Carboxylic acids (C, H, O and halogen, N or S) (*cont.*)

| | M.p. | Amide (p. 32) | Anilide (p. 32) | 4-Toluidide (p. 32) | 4-Bromophenacyl ester (p. 32) | 4-Phenylphenacyl ester (p. 32) | Notes |
|---|---|---|---|---|---|---|---|
| 3-Chlorobenzoic | 158 | 134 | 124 | | 117 | 154 | |
| 2,4-Dichlorobenzoic | 160 | 194 | 165 (153) | | | | Conc. $HNO_3$ + $H_2SO_4$ → 3,5-dinitro deriv., 212 |
| 2-Iodobenzoic | 162 | 184 | 142 | | 110 | 143 | |
| 5-Bromo-2-hydroxybenzoic | 165 | 232 | 222 | | | | Ethanoate, 168 (p. 41) |
| 4-Nitrophthalic | 165 | 200d | 192 | 172 | | 120 | |
| 2-Mercaptobenzoic | 165 | | | | | | Ethanoate, 125 (p. 41); See Tables 9 and 13 |
| 3-Aminobenzoic | 174 | 111 | 140 | | 190* | | *Two moles of reagent are required to form this derivative |
| 4-Chloro-3-nitrobenzoic | 181 | 156 | 131 | | | | |
| 4-Fluorobenzoic | 182 | 154 | | | | | |
| 2,4-Dinitrobenzoic | 183 | 203 | | | 158 | | |
| 4-Aminobenzoic | 186 | 114 | | | 200* | | See Tables 9 and 13; *See under 3-aminobenzoic acid |
| 3-Iodobenzoic | 187 | 186 | | | 128 | 147 | |
| N-Benzoylglycine (Hippuric) | 187 | 183 | 208 | | 151 | 163 | |
| 3,5-Dinitrobenzoic | 204 | 183 | 235 | 280 | 159 | 154 | |
| 3-Nitrophthalic | 218 | 201d | 234 | 224 | 166 | 149 | Acyl chloride, 70; Anhydride, 162 |
| 2-Hydroxy-5-nitrobenzoic | 229 | 225 | 224 | | | | |
| Pyridine-3-carboxylic (Nicotinic) | 237 | 122 | 132* | 150 | | | *From benzene; from water, 85 |
| 3-(2-Nitrophenyl)propenoic (2-Nitrocinnamic) | 239 | 185 | | | 142 | 146 | Unsaturated |
| 4-Nitrobenzoic | 240 | 201 | 211 | 203 | 134 | 182 | Acyl chloride, 75 |
| 4-Chlorobenzoic | 241 | 179 | 194 | | 126 | 160 | |
| 4-Bromobenzoic | 251 | 189 | 197 | | 134 | 160 | |
| 4-Iodobenzoic | 265 | 217 | 210 | | 146 | 171 | |
| 3-(4-Nitrophenyl)propenoic (4-Nitrocinnamic) | 285 | 217 | | | 191 | 192 | Unsaturated |
| Pyridine-4-carboxylic (Isonicotinic) | 324* | 155 | | | | | *Sublimes |

**Table 18.** Enols

| | B.p. | Semi-carbazone (p. 33) | 2,4-Dinitro-phenyl-hydrazone (p. 33) | Colour with aq. FeCl₃ | Notes |
|---|---|---|---|---|---|
| Pentan-2,4-dione (Acetylacetone) | 139 | 107* | 122* 209 di | Red | *Pyrazole deriv. See Table 26 |
| Methyl 3-oxobutanoate (Methyl acetoacetate) | 170 | 152 | 119 | Red | See Table 26 |
| Ethyl 3-oxobutanoate (Ethyl acetoacetate) | 181d | 133 | 96 | Red | See Table 26 |
| Ethyl 3-oxopentanedioate (Ethyl acetonedicarboxylate) | 250 | 94 | 86 | Red | See Table 26 |
| | M.p. | | | | |
| 1-Phenylbutan-1,3-dione (Benzoylacetone) | 61 | | 151 | Red | See Table 26 |
| 1,3,5-Trihydroxybenzene (Phloroglucinol) | 217 | | | Violet* | *Transient colour. Picrate, 101. See Table 30 |

**Table 19.** Esters, carboxylic

| | B.p. | Notes |
|---|---|---|
| Methyl methanoate | 32 | |
| Ethyl methanoate | 54 | |
| Methyl ethanoate | 57 | |
| Isopropyl methanoate | 68 | |
| Vinyl ethanoate | 72 | Unsaturated; polymerizes readily |
| Methyl chloromethanoate | 73 | Very reactive chlorine atom |
| Ethyl ethanoate | 77 | |
| Methyl propanoate | 79 | |
| Propyl methanoate | 81 | |
| Prop-2-enyl methanoate | 83 | Unsaturated |
| Methyl propenoate | 85 | Unsaturated; polymerizes readily |
| Isopropyl ethanoate | 91 | |
| Methyl 2-methylpropanoate | 92 | |
| Ethyl chloromethanoate | 93 | Very reactive chlorine atom |
| But-2-yl methanoate | 97 | |
| 2-Methylpropyl methanoate | 98 | |
| Ethyl propanoate | 98 | |
| 2-Methylprop-2-yl ethanoate | 98 | |
| Methyl 2-methylpropenoate | 99 | Unsaturated; polymerizes readily |
| Propyl ethanoate | 101 | |
| Ethyl propenoate | 101 | Unsaturated; polymerizes readily |
| Methyl butanoate | 102 | |
| Prop-2-enyl ethanoate | 103 | Unsaturated |
| Isopropyl chloromethanoate | 105 | Very reactive chlorine atom |
| Butyl methanoate | 107 | |
| Ethyl 2-methylpropanoate | 110 | |
| But-2-yl ethanoate | 111 | |
| Isopropyl propanoate | 111 | |
| Isobutyl ethanoate | 116 | |
| Methyl 3-methylbutanoate | 116 | |
| Ethyl butanoate | 120 | |
| Propyl propanoate | 122 | |

## Table 19. Esters, carboxylic (*cont.*)

| | B.p. | Notes |
|---|---|---|
| 3-Methylbutyl methanoate | 123 | |
| Butyl ethanoate | 125 | |
| Isopropyl butanoate | 128 | |
| Methyl pentanoate | 130 | |
| 2-Methylpropyl chloromethanoate | 130 | Very reactive chlorine atom |
| Methyl chloroethanoate | 130 | Very reactive chlorine atom |
| Ethyl 3-methylbutanoate | 134 | |
| Methyl 2-oxopropanoate | 136 | 2,4-Dinitrophenylhydrazone, 187 (p. 27). See Table 26 |
| 2-Methylpropyl propanoate | 137 | |
| Ethyl but-2-enoate | 138 | Unsaturated |
| 3-Methylbutyl ethanoate | 142 | |
| Propyl butanoate | 143 | |
| Butyl propanoate | 144 | |
| Methyl 2-hydroxypropanoate | 144 | |
| Methyl bromoethanoate | 144d | Very reactive bromine atom |
| Ethyl pentanoate | 145 | |
| Ethyl chloroethanoate | 145 | Very reactive chlorine atom |
| Ethyl 2-chloropropanoate | 146 | Very reactive chlorine atom |
| Benzyl chloroethanoate | 147 | Very reactive chlorine atom |
| Methyl hexanoate | 150 | |
| Ethyl 2-hydroxypropanoate | 152 | |
| Ethyl 2-oxopropanoate | 155 | Semicarbazone, 206d (p. 27); 2,4-dinitrophenyl-hydrazone, 155 (p. 27) |
| 2-Methylpropyl butanoate | 157 | |
| Ethyl bromoethanoate | 159 | Very reactive bromine atom |
| Ethyl hydroxyethanoate | 160 | |
| Ethyl 2-bromopropanoate | 162 | Very reactive bromine atom |
| Ethyl hexanoate | 166 | |
| Ethyl trichloroethanoate | 167 | |
| Methyl 3-oxobutanoate | 170 | Red colour with aq. FeCl$_3$; semicarbazone, 152 (p. 27); see Table 26 |
| Methyl heptanoate | 173 | |
| Ethyl 3-bromopropanoate | 179 | |
| Methyl 2-furoate | 181 | |
| Dimethyl propanedioate | 181 | |
| Ethyl 3-oxobutanoate | 181 | Red colour with aq. FeCl$_3$; semicarbazone, 129 (p. 27); see Table 26 |
| Diethyl ethanedioate | 186 | |
| Ethyl heptanoate | 189 | |
| Methyl octanoate | 193 | |
| Tetrahydrofurfuryl ethanoate | 194 | |
| Dimethyl butanedioate | 195 | |
| Methyl 4-oxopentanoate | 196 | See Table 26 |
| Phenyl ethanoate | 196 | |
| Methyl benzoate | 198 | |
| Diethyl propanedioate | 199 | |
| Methyl cyanoethanoate | 200 | |
| Benzyl methanoate | 203 | |
| γ-Butyrolactone | 204 | |
| Ethyl 4-oxopentanoate | 205 | Semicarbazone, 135 (p. 27) |
| Dimethyl *cis*-butenedioate | 205 | Unsaturated |
| Ethyl octanoate | 206 | |
| γ-Pentanolactone | 207 | |
| 2-Methylphenyl ethanoate | 208 | |
| Phenyl propanoate | 211 | |
| Ethyl benzoate | 212 | |

**Table 19.** Esters, carboxylic (*cont.*)

| | B.p. | Notes |
|---|---|---|
| 3-Methylphenyl ethanoate | 212 | |
| 4-Methylphenyl ethanoate | 212 | |
| Methyl 2-toluate | 213 | |
| Benzyl ethanoate | 214 | |
| Methyl 3-toluate | 215 | |
| Diethyl *trans*-butenedioate | 216 | Unsaturated |
| Diethyl butanedioate | 216 | |
| Methyl 4-toluate (m.p. 33) | 217 | |
| Methyl phenylethanoate | 220 | |
| Methyl 2-hydroxybenzoate | 224 | See Table 30 |
| Diethyl *cis*-butenedioate | 225 | Unsaturated |
| Phenyl butanoate | 227 | |
| (−)-Menthyl ethanoate | 227 | |
| Ethyl phenylethanoate | 227 | |
| Prop-2-enyl benzoate | 230 | Unsaturated |
| Ethyl 2-hydroxybenzoate | 233 | See Table 30 |
| Benzyl butanoate | 238 | |
| Dibutyl ethanedioate | 243 | |
| Methyl 2-bromobenzoate | 244 | |
| Diethyl hexanedioate | 245 | |
| Dipropyl butanedioate | 246 | |
| Methyl undecyl-10-enoate | 248 | Unsaturated |
| Butyl benzoate | 249 | |
| Diethyl 3-oxopentanedioate | 250 | Red colour with aq. FeCl$_3$; semicarbazone, 94 (p. 27); see Table 26 |
| Ethyl 2-hydroxy-2-phenylethanoate (m.p. 37) | 254 | |
| Methyl 2-aminobenzoate | 255d | See Table 9 |
| Glyceryl triethanoate | 258 | |
| Methyl dodecanoate | 258 | |
| Butyl 2-hydroxybenzoate | 259 | See Table 30 |
| Glyceryl monoethanoate | 260 | |
| 2-Methylpropyl 2-hydroxybenzoate | 261 | See Table 30 |
| Methyl 3-phenylpropenoate (m.p. 33) | 263 | Unsaturated |
| Ethyl 2-aminobenzoate | 265d | See Table 9 |
| Ethyl benzoylethanoate | 265 | See Tables 18 and 26 |
| Ethyl 4-methoxybenzoate | 269 | |
| Ethyl 3-phenylpropenoate (m.p. 12) | 271 | Unsaturated |
| Diethyl dihydroxybutanedioate (m.p. 17) | 280 | |
| Dimethyl benzene-1,2-dicarboxylate | 282 | |
| Dimethyl decanedioate (m.p. 38) | 290 | |
| Diethyl benzene-1,2-dicarboxylate | 298 | |
| Diethyl benzene-1,3-dicarboxylate | 302 | |
| Ethyl 2-naphthoate (m.p. 32) | 308 | |
| Ethyl 1-naphthoate | 309 | |
| Benzyl benzoate | 323 | |
| Dibutyl benzene-1,2-dicarboxylate | 338 | |
| Methyl octadecanoate (m.p. 38) | 214/15 mm | |

| | M.p. | |
|---|---|---|
| Glyceryl 1,3-diethanoate | 40 | |
| Dibenzyl benzene-1,2-dicarboxylate | 42 | |
| Phenyl 2-hydroxybenzoate | 42 | See Table 30 |
| Diethyl benzene-1,4-dicarboxylate | 43 | |
| Methyl 4-chlorobenzoate | 44 | |
| Dibenzyl butanedioate | 45 | |
| Methyl 4-methoxybenzoate | 45 | |

**Table 19.** Esters, carboxylic (*cont.*)

| | B.p. | Notes |
|---|---|---|
| Ethyl 3-nitrobenzoate | 47 | |
| (+)-Dimethyl dihydroxybutanedioate | 48 | |
| 1-Naphthyl ethanoate | 49 | |
| Dimethyl ethanedioate | 52 | |
| Ethyl 4-nitrobenzoate | 56 | |
| Methyl 2-hydroxy-2-phenylethanoate | 58 | Benzoate, 76 (p. 26) |
| Dimethyl benzene-1,3-dicarboxylate | 67 | |
| Phenyl benzoate | 68 | |
| Methyl 3-hydroxybenzoate | 70 | Violet colour with aq. $FeCl_3$ |
| 2-Naphthyl ethanoate | 70 | |
| Glyceryl tristearate | 71 | |
| Phenyl 3-phenylpropenoate | 72 | Unsaturated |
| Ethyl 3-hydroxybenzoate | 72 | Violet colour with aq. $FeCl_3$. Benzoate, 58 (p. 41) |
| Methyl 3-nitrobenzoate | 78 | |
| Dibenzyl ethanedioate | 80 | |
| Ethyl 4-aminobenzoate | 90 | See Table 9 |
| Methyl 4-nitrobenzoate | 96 | |
| Dimethyl *trans*-butenedioate | 102 | Unsaturated |
| Ethyl 4-hydroxybenzoate | 115 | Violet colour with aq. $FeCl_3$. See Table 30. Benzoate 89 (p. 41) |
| Methyl 4-hydroxybenzoate | 131 | Violet colour with aq. $FeCl_3$. See Table 30. Benzoate, 135. (p. 41) |
| Dimethyl benzene-1,4-dicarboxylate | 140 | |
| Methyl 3-(4-nitrophenyl)propenoate | 161 | Unsaturated |

**Table 20.** Esters, phosphoric. Hydrolysis of the ester (p. 33) gives an alcohol or phenol which should be identified in the usual way

| | B.p |
|---|---|
| Trimethyl phosphate | 197 |
| Triethyl phosphate | 215 |
| Tripropyl phosphate | 138/47 mm |
| Tri-2-tolyl phosphate | 265/20 mm |
| Tributyl phosphate | 157/10 mm |

| | M.p. |
|---|---|
| Triphenyl phosphate | 50 |
| Tribenzyl phosphate | 64 |
| Tri-4-tolyl phosphate | 78 |
| Tri-2-phenylphenyl phosphate | 113 |

**Table 21.** Ethers

| | B.p. | M.p. | Alkyl 3,5-di-nitro-ben-zoate (p. 35) | Picric acid com-plex (p. 35) | Sulphon-amide (p. 35) posi-tion | M.p. | Nitro deriv. (p. 34) posi-tion | M.p. | me-thod | Notes |
|---|---|---|---|---|---|---|---|---|---|---|
| Diethyl ether | 35 | | 93 | | | | | | | |
| Chloromethyl methyl ether | 59 | | | 163 | | | | | | |
| Tetrahydrofuran | 65 | | | | | | | | | |
| Di-isopropyl ether | 68 | | 122 | | | | | | | |
| Dipropyl ether | 90 | | 75 | | | | | | | |
| Di-(prop-2-enyl) ether | 94 | | 50 | | | | | | | Unsaturated |
| Dioxan | 101 | 11 | | | | | | | | Miscible with water |
| 1-Chloro-2,3-epoxypropane (Epichlorohydrin) | 116 | | | | | | | | | Boiling ethanolic NaOH + phenol → glycerol di-phenyl ether 81 |
| Di-(1-methylpropyl) ether | 121 | | 76 | | | | | | | |
| Di-(2-methylpropyl) ether | 122 | | 88 | | | | | | | |
| 2-Ethoxyethanol (Ethylene glycol monoethyl ether) | 135 | | 75 | | | | | | | Miscible with water |
| Dibutyl ether | 142 | | 64 | | | | | | | |
| Methoxybenzene (Anisole) | 154 | | | | 4 | 111 | 2,4 | 94* 87* | i | *Two crystalline forms |
| 2-Methoxytoluene | 171 | | 69 | 117 | 5 | 137 | 3,5 | 69 | iv | |
| Ethoxybenzene (Phenetole) | 172 | | | 92 | 4 | 149 | 4 | 59 | iii | |
| Di-(3-methylbutyl) ether | 172 | | 62 | | | | | | | |
| Benzyl methyl ether | 172 | | | 116 | | | | | | |
| 4-Methoxytoluene | 176 | | | 89 | 3 | 182 | 3,5 | 122 | ii | |
| 3-Methoxytoluene | 177 | | | 114 | 6 | 130 | 2 | 54 | v | |
| | | | | | | | 2,4,6 | 94 | i | |
| Dipenyl ether | 187 | | 46 | | | | | | | |
| 4-Ethotoluene | 190 | | | 111 | 3 | 138 | 3,5 | 71 | i | |
| 3-Ethtoluene | 192 | | | 115 | 6 | 110 | 6 | 54 | v | |
| 2-Chmethoxy-zene | 195 | | | | 4 | 130 | 4 | 95 | iii | |
| romethoxy-nzene (p-Chloroanisole) | 198 | | | | 2 | 151 | 2 | 98 | iii | |
| -Dimethoxy-benzene(Veratrole) | 207 | 22 | | 56 | 4 | 136 | 4 | 95 | v | |
| | | | | | | | 4,5 | 132 | ii | |
| 2-Chlorophenyl ethyl ether (o-Chlorophene-tole) | 208 | 17 | | | 4 | 133 | 4 | 82 | iii | |
| Butoxybenzene | 210 | | | 112 | 4 | 101 | | | | |

**Table 21.**  Ethers (*cont.*)

| | B.p. | M.p. | Alkyl 3,5-di-nitro-ben-zoate (p. 35) | Picric acid com-plex (p. 35) | Sulphon-amide (p. 35) posi-tion | M.p. | Nitro deriv. (p. 34) posi-tion | M.p. | me-thod | Notes |
|---|---|---|---|---|---|---|---|---|---|---|
| 4-Chlorophenyl ethyl ether (*p*-Chlorophene-tole) | 212 | 21 | | | 2 | 134 | 2,6 | 54 | iii | |
| 1,3-Dimethoxy-benzene | 217 | | | 57 | 4 | 166 | 2,4,6 | 124 | ii | |
| 2-Bromophenyl ethyl ether (*o*-Bromophene-tole) | 218 | | | | 4 | 134 | 4 | 98 | iii | |
| 2-Bromomethoxy-benzene (*o*-Bromoanisole) | 218 | | | | 4 | 140 | 4 | 106 | iii | |
| 4-Bromomethoxy-benzene (*p*-Bromoanisole) | 223 | 12 | | | 2 | 148 | 2 | 88 | iii | |
| Dihexyl ether | 228 | | | | | | | | | |
| 4-Bromophenyl ethyl ether (*p*-Bromophene-tole) | 229 | 12 | 61 | | 2 | 144 | 2 | 47 | iii | |
| 1,2-Methylenedioxy-4-(prop-2-enyl)-benzene (Safrole) | 232 | 11 | | 104 | | | 1,3,5 | 51 | iii | Unsaturated; $Br_2$ in ether → pentabromo deriv., 169 (p. 35) |
| 1,3-Diethoxybenzene | 235 | 12 | | 109 | 4 | 184 | | | | $Br_2$ in ethanoic acid → tribromo deriv., 69 (p. 35) |
| (4-Prop-1-enyl)meth-oxybenzene (Anethole) | 235 | 21 | | 70 | | | | | | Unsaturated: tri-bromo deriv., 108 (p. 35) $CrO_3$-ethanoic acid → 4-methoxy-benzoic acid, 184 (p. 27) |
| Diphenyl ether | 259 | 28 | | 110 | 4,4′ | 159 | 4,4′ | 144 | iii | |
| 1-Methoxynaph-thalene | 271 | | | 130 | 4 | 156 | 2,4,5 | 128 | iii | |
| 1-Ethoxynaph-thalene | 280 | 5 | | 119 | 4 | 164 | 2,4,5 | 149 | iii | |
| 2-Ethoxynaph-thalene | 282 | 37 | | 101 | 8 | 162 | 1,6,8 | 186 | iii | |
| Dibenzyl ether | 295d | 3 | 112 | 78 | | | | | | Dibromo deriv., 107 (p. 35) |

**Table 21.** Ethers (*cont.*)

| | B.p. | M.p. | Alkyl 3,5-di-nitro-ben-zoate (p. 35) | Picric acid com-plex (p. 35) | Sulphon-amide (p. 35) posi-tion | | Nitro deriv. (p. 34) posi-tion | M.p. | me-thod | Notes |
|---|---|---|---|---|---|---|---|---|---|---|
| 1,2-Diethoxybenzene | 217 | 43 | | 71 | 4 | 162 | 4 | 73 | v | |
| | | | | | | | 3,4,6 | 122 | ii | |
| 1,2,3-Trimethoxy-benzene | 241 | 47 | | 81 | 2,3,4 | 123 | 5 | 106 | ii | |
| 1,4-Dimethoxy-benzene | 212 | 56 | | 48 | 2 | 148 | 2 | 72 | i | |
| 1,4-Diethoxybenzene | | 72 | | | 2 | 154 | 2 | 49 | i | |
| 2-Methoxynaphthalene | | 72 | | 117 | 8 | 151 | 1,6,8 | 215d | iii | |
| 1,4-Dibenzyloxybenzene | | 127 | | | | | 2 | 83 | i | |

**Table 22.** Halides, alkyl mono-

| | Chlo-ride B.p. | Bro-mide B.p. | Io-dide B.p. | Thio-uro-nium pic-rate (p. 36) M.p. | 2-Naph-thyl ether (p. 36) M.p. | 2-Naph-thyl ether picrate (p. 36) M.p. | Notes |
|---|---|---|---|---|---|---|---|
| Methyl | −24 | 4 | 43 | 224 | 72 | 118 | |
| Ethyl | 12 | 38 | 72 | 188 | 37 | 102 | |
| Prop-2-yl | 36 | 60 | 89 | 196 | 41 | 95 | |
| Prop-2-enyl | 46 | 70 | 102 | 154 | 16 | 99 | Unsaturated |
| Propyl | 46 | 71 | 102 | 177 | 40 | 81 | |
| 2-Methylprop-2-yl(t-Butyl) | 51 | 72 | 98 | 151 | | | |
| 1-Methylpropyl(s-Butyl) | 67 | 90 | 120 | 166 | 34 | 86 | |
| 2-Methylpropyl(isobutyl) | 68 | 91 | 120 | 167 | 33 | 84 | |
| Butyl | 77 | 101 | 130 | 177 | 33 | 67 | |
| 3-Methylbutyl (Isopentyl) | 100 | 119 | 147 | 173 | 28 | 94 | |
| Pentyl | 107 | 128 | 155 | 154 | 25 | 66 | |
| 1-Chloro-2,3-epoxypropane (Epichlorohydrin) | 116 | | | | | | See Table 21 |
| Hexyl | 134 | 156 | 180 | 157 | | | |
| Cyclohexyl | 142 | 166 | 180d | | 116 | | |
| Heptyl | 160 | 178 | 203 | 142 | | | |
| Benzyl | 179 | 198 | 24* | 187 | 102 | 122 | *M.p. |
| Octyl | 183 | 203 | 225 | 134 | | | |
| 2-Phenylethyl | 190 | 218 | | 139 | 70 | 83 | |
| 1-Phenylethyl | 195 | 205 | | 167 | | | |
| 2-Chlorobenzyl | 214 | 102/9 mm | | 213 | | | $CrO_3 \rightarrow$ 2-chlorobenzoic acid, 142 (p. 27) |
| 3-Chlorobenzyl | 215 | 109/10 mm | | 200 | | | $CrO_3 \rightarrow$ 3-chlorobenzoic acid, 158 |
| 2-Bromobenzyl | 125/20 mm | 31* | 47* | 222 | | | *M.p. $CrO_3 \rightarrow$ 2-bromo-benzoic acid, 150 |

**Table 22.** Halides, alkyl mono- (*cont.*)

| | Chloride B.p. | Bromide B.p. | Iodide B.p. | Thiouronium picrate (p. 36) M.p. | 2-Naphthyl ether (p. 36) M.p. | 2-Naphthyl ether picrate (p. 36) M.p. | Notes |
|---|---|---|---|---|---|---|---|
| 3-Bromobenzyl | 119/18 mm | 41* | 42* | 205 | | | *M.p. CrO₃ → 3-bromobenzoic acid, 155 |
| 4-Chlorobenzyl | 214 | 51* | 64* | 194 | | | *M.p. CrO₃ → 4-chlorobenzoic acid, 241 |
| | M.p. | M.p. | M.p. | | | | |
| 3-Nitrobenzyl | 45 | 58 | 84 | | | | KMnO₄ → 3-nitrobenzoic acid, 141 |
| 2-Nitrobenzyl | 48 | 46 | 75 | | | | KMnO₄ → 2-nitrobenzoic acid, 146 |
| 4-Bromobenzyl | 50 | 63 | 80 | 219 | | | CrO₃ → 4-bromobenzoic acid, 251 |
| 4-Nitrobenzyl | 71 | 99 | 127 | | | | KMnO₄ → 4-nitrobenzoic acid, 240 |

**Table 23.** Halides, alkyl poly-

| | B.p. | Thiouronium picrate (p. 37) | 2-Naphthyl ether (p. 37) | Notes |
|---|---|---|---|---|
| Dichloromethane (Methylene dichloride) | 42 | 267 | 133 | |
| *trans*-1,2-Dichloroethene | 48 | | | Unsaturated. Dibromide, 192 |
| *cis*-1,2-Dichloroethene | 60 | | | Unsaturated. Dibromide, 192 |
| 1,1-Dichloroethane (Ethylidene dichloride) | 60 | | 200 | |
| Trichloromethane (Chloroform) | 61 | | | Forms carbylamine with pri.amines and boiling ethanolic KOH |
| Carbon tetrachloride | 77 | | | Forms carbylamine (as above) |
| 1,2-Dichloroethane | 83 | 260 | 217 | |
| Trichloroethene | 90 | | | Unsaturated. Chlorine atoms are unreactive |
| Dibromomethane | 97 | 267 | 133 | |
| 1,2-Dichloropropane | 98 | 232 | 152 | |
| 1-Bromo-2-chloroethane | 106 | | 217 | |
| 1,1-Dibromoethane | 112 | | | |
| Tetrachloroethene | 121 | | 117 | Unsaturated. Chlorine atoms are unreactive |
| 1,3-Dichloropropane | 123 | 229 | 148 | |
| 1,2-Dibromoethane (m.p. 10) | 132 | 260 | 217 | |
| 1,2-Dibromopropane | 141 | 232 | 152 | |
| 1,1,2,2-Tetrachloroethane | 147 | | | |

**Table 23.** Halides, alkyl poly- (*cont.*)

| | B.p. | Thio-uronium picrate (p. 37) | 2-Naph-thyl ether (p. 37) | Notes |
|---|---|---|---|---|
| Tribromomethane (Bromoform) | 149 | | | Forms carbylamine (see chloroform) |
| 1,3-Dibromopropane | 167 | 229 | 148 | |
| Di-iodomethane | 180 | | 133 | |
| (Dichloromethyl)benzene (Benzal chloride) | 212 | | | Conc. $H_2SO_4$ at 50° → benzaldehyde, which may be characterized as 2,4-dinitrophenylhydrazone, 237 |
| Trichloromethylbenzene (Benzotrichloride) | 221 | | | Boiling with aq. $Na_2CO_3$ → benzoic acid, 122 |
| 1,5-Dibromopentane | 221 | 247 | | |
| 1,3-Di-iodopropane | 224 | | 148 | |
| | M.p. | | | |
| 1,2-Di-iodoethane | 82 | 260 | 217 | |
| Carbon tetrabromide | 91 | | | |
| Tri-iodomethane (Iodoform) | 119 | | | Yellow. With quinoline in ether → compound, 65 |
| Hexachloroethane | 187* | | | *Sublimes. Camphor-like odour |

**Table 24.** Halides, aryl

| | B.p. | M.p. | Sulph-onamide (p. 37) position | M.p. | Nitro deriv. (p. 37) position | M.p. | method | Notes |
|---|---|---|---|---|---|---|---|---|
| Fluorobenzene | 85 | | 4 | 125 | 4 | 27 | ii | |
| 2-Fluorotoluence | 114 | | 5 | 105 | | | | |
| 3-Fluorotoluene | 116 | | 6 | 173 | | | | |
| 4-Fluorotoluene | 117 | | 2 | 141 | | | | |
| Chlorobenzene | 132 | | 4 | 143 | 2,4 | 52 | ii | |
| Bromobenzene | 156 | | 4 | 162 | 2,4 | 75 | ii | |
| 2-Chlorotoluene | 159 | | 5 | 126 | 3,5 | 64 | ii | |
| 3-Chlorotoluene | 162 | | 6 | 185 | 4,6 | 91 | ii | |
| 4-Chlorotoluene | 162 | | 2 | 143 | 2,6 | 76 | i | |
| 1,3-Dichlorobenzene | 173 | | 4 | 180 | 4,6 | 103 | ii | |
| 1,2-Dichlorobenzene | 179 | | 4 | 135 | 4,5 | 110 | ii | |
| 2-Bromotoluene | 181 | | 5 | 146 | 3,5 | 82 | ii | |
| 3-Bromotoluene | 184 | | 6 | 168 | 4,6 | 103 | i | |
| 4-Bromotoluene | 185 | 28 | 2 | 165 | 2 | 47 | iii | |
| Iodobenzene | 188 | | | | 4 | 174 | i | $Br_2$ + Fe → 4-bromo deriv., 91 |
| 2,4-Dichlorotoluene | 197 | | 5 | 176 | 3,5 | 104 | ii | |
| 2,6-Dichlorotoluene | 199 | | 3 | 204 | 3,5 | 121 | ii | |
| 3-Iodotoluene | 204 | | | | 6 | 84 | i | |
| 2-Iodotoluene | 211 | | | | 6 | 103 | ii | |
| 4-Iodotoluene | 211 | 35 | | | | | | Boiling $HNO_3 \xrightarrow{3 \text{ hr}}$ acid, 265 |
| 2-Chlorobenzyl chloride | 214 | | | | | | | See Table 22 |
| 4-Chlorobenzyl chloride | 214 | 29 | | | | | | See Table 22 |
| 1,2,4-Trichlorobenzene | 214 | 17 | 5 | > 200 | 3,5 | 103 | ii | |
| 1-Fluoronaphthalene | 214 | | | | | | | Picric acid deriv., 113 (p. 37) |

Table 24. Halides, aryl (*cont.*)

| | B.p. | M.p. | Sulph-onamide (p. 37) position | M.p. | Nitro deriv. (p. 37) position | M.p. | method | Notes |
|---|---|---|---|---|---|---|---|---|
| 3-Chlorobenzyl chloride | 215 | | | | | | | See Table 22 |
| 1,3-Dibromobenzene | 219 | | 4 | 190 | 4,6 | 117 | iii | |
| 1,2-Dibromobenzene | 224 | | 4 | 175 | 4,5 | 114 | ii | |
| 1-Chloronaphthalene | 260 | | 4 | 186 | 4,5 | 180 | iii | |
| 1-Bromonaphthalene | 281 | | 4 | 192 | 4 | 85 | iii | Picric acid deriv., 134 |
| 1-Iodonaphthalene | 305 | | | | | | | Picric acid deriv., 128 |
| 3-Bromobenzyl bromide | | 41 | | | | | | See Table 22 |
| 1,4-Dichlorobenzene | 174 | 53 | 2 | 180 | 2 | 56 | iii* | *Without cooling |
| 1,2,3-Trichlorobenzene | 218 | 53 | 4 | 227 | 4 | 56 | ii | |
| 2-Iodonaphthalene | | 55 | | | | | | Picric acid deriv., 95 (p. 37) |
| 2-Bromonaphthalene | | 59 | 8 | 208 | | | | Picric acid deriv., 86 (p. 37) |
| 2-Chloronaphthalene | | 60 | 8 | 232 | 1,8 | 175 | ii* | *6 hr. at 100° |
| 4-Bromobenzyl bromide | | 63 | | | | | | See Table 22 |
| 1,3,5-Trichlorobenzene | | 63 | 2 | 212d | 2 | 68 | ii | |
| 4-Bromochlorobenzene | | 67 | | | 2 | 72 | iii* | *Without cooling |
| 1,4-Dibromobenzene | | 87 | 2 | 195 | 2 | 84 | iii* | *Without cooling |
| 1,3,5-Tribromobenzene | | 120 | 2 | 222d | 2,4 | 192 | ii | |
| 1,2,3,4-Tetrachlorobenzene | | 139 | 3 | 99 | | | | |
| | | | 3,6 | 227 | | | | |

Table 25. Hydrocarbons

| | B.p. | M.p. | Sulph-ona-mide (p. 38) | Nitro deriv. (p. 34) position | M.p. | meth-od | Piric acid deriv. (p. 29) | Notes |
|---|---|---|---|---|---|---|---|---|
| 2-Methyl-1,3-butadiene (Isoprene) | 34 | | | | | | | Unsaturated; polymerizes easily; maleic anhydride adduct, 64 (p. 38) |
| Pent-1-yne | 40 | | | | | | | Unsaturated. Hg deriv., 118 (pp. 38) |
| Cyclopentadiene | 40 | | | | | | | Unsaturated. Maleic anhydride adduct, 164 (p. 38) Forms dimer b.p. 170d, m.p. 32 on standing |
| Penta-1,3-diene (Piperylene) | 42 | | | | | | | Unsaturated. Maleic anhydride adduct, 61 (p. 38) |
| Cyclopentene | 44 | | | | | | | Unsaturated |
| Benzene | 80 | 6 | 153 | 1,3 | 90 | ii | | |
| Cyclohexane | 81 | 6 | | | | | | Oxidation with fuming HNO$_3$ → hexanedioic acid, 153 |
| Cyclohexene | 83 | | | | | | | Unsaturated. Conc. HNO$_3$ → hexanedioic acid, 153 |
| Toluene | 110 | | 137 | 2,4 | 70 | ii | | |
| Ethylbenzene | 136 | | 109 | 2,4,6 | 37 | ii | | |
| 1,4-Dimethylbenzene (p-Xylene) | 137 | 15 | 147 | 2,3,5 | 139 | ii | | |

**Table 25.** Hydrocarbons (*cont.*)

| | B.p. | M.p. | Sulphona-mide (p.38) | Nitro deriv. (p. 34) position | Nitro deriv. (p. 34) M.p. | method | Piric acid deriv. (p. 29) | Notes |
|---|---|---|---|---|---|---|---|---|
| 1,3,-Dimethylbenzene (*m*-Xylene) | 139 | | 137 | 2,4,6 | 182 | ii | | |
| Phenylethyne | 140 | | | | | | | Unsaturated. Hg deriv., 125 (pp. 38) |
| 1,2-Dimethylbenzene (*o*-Xylene) | 144 | | 144 | 4,5 | 71 | ii | | |
| Phenylethene (Vinylbenzene, styrene) | 146 | | | | | | | Unsaturated; polymerizes in the presence of a drop of $H_2SO_4$. Dibromide, 73 |
| Isopropylbenzene (Cumene) | 153 | | 105 | 2,4,6 | 109 | ii | | |
| α-Pinene | 156 | | | | | | | Unsaturated. Dibromide, 164 |
| Allylbenzene | 157 | | | | | | | Unsaturated. $CrO_3 \rightarrow$ benzoic acid, 122 |
| Propylbenzene | 159 | | 107 | | | | | |
| 1,3,5-Trimethylbenzene (Mesitylene) | 165 | | 142 | 2,4,6 | 235 | i | | |
| 1,2,4-Trimethylbenzene (Pseudocumene) | 168 | | 181 | 3,5,6 | 185 | ii | | |
| Dicyclopentadiene | 170d | 32 | | | | | | Unsaturated. Benzoquinone adduct, 157 (p. 38) |
| (+)-Limonene | 176 | | | | | | | Unsaturated; odour of lemons. Tetrabromide 104 |
| 4-(Prop-2-yl)toluene (*p*-Cymene) | 176 | | 115 | 2,3,6 | 118 | ii | | |
| Dipentene (Limonene) | 181 | | | | | | | Unsaturated; odour of lemons. Tetrabromide, 124 |
| Indene | 182 | | | | | | | Unsaturated; polymerized by acid or heat. $HNO_3 \rightarrow$ phthalic acid, 195 Benzylidene deriv., 135 (p. 39) |
| Tetrahydronaphthalene (Tetralin) | 207 | | 135 | 5,7 | 95 | i | | |
| 1-Methylnaphthalene | 241 | | | 4 | 71 | iii | 141 | |
| 2-Methylnaphthalene | 241 | 37 | | 1 | 81 | i | 115 | |
| Diphenylmethane | 262 | 26 | | 2,2',4,4' | 172 | ii | | $CrO_3 - H_2SO_4 \rightarrow$ benzophenone, 48 |
| (−)-Camphene | 160 | 51 | | | | | | Unsaturated; dibromide, 89 |
| 1,2-Diphenylethane (Dibenzyl) | 284 | 52 | | 4,4' | 180 | i | | $CrO_3 - H_2SO_4 \rightarrow$ benzoic acid, 122 |
| Biphenyl | 255 | 70 | | 4,4' | 234 | iv | | $Br_2$-ethanoic acid (boil for 2 hr) $\rightarrow$ 4,4'-dibromo deriv., 169 |
| 1,2,4,5-Tetramethylbenzene (Durene) | | 79 | 155 | 3,6 | 205 | ii | 95 | |
| Naphthalene | | 80 | | 1 | 61 | i | 150 | Odour of 'moth balls'; styphnic acid deriv., 168 |
| Acenaphthylene | | 92 | | | | | 201 | Dibromide, 121 |
| Triphenylmethane | | 94 | | 4,4',4" | 206 | iii | | |
| Acenaphthene | | 95 | | 5 | 101 | i | 161 | Styphnic acid deriv., 154 |
| Phenanthrene | | 100 | | | | | 144 | $CrO_3$-ethanoic acid $\rightarrow$ quinone, 202. Styphnic acid deriv., 142 |

### Table 25. Hydrocarbons (*cont.*)

| | B.p. | M.p. | Sulph-ona-mide (p. 38) | Nitro deriv. (p. 34) posi-tion | M.p. | meth-od | Picric acid deriv. (p. 29) | Notes |
|---|---|---|---|---|---|---|---|---|
| 2,3-Dimethylnaphthalene | | 104 | | | | | 124 | Styphnic acid deriv., 149 |
| Fluoranthene | | 110 | | | | | 182 | Styphnic acid deriv., 151 |
| 2,6-Dimethylnaphthalene | | 111 | | | | | 143 | Styphnic acid deriv., 159 |
| Fluorene | | 115 | | 2 | 156 | i | 84* | Gives blue colour with conc. |
| | | | | 2,7 | 199 | ii | | $H_2SO_4$; styphnic acid deriv., 134 |
| | | | | | | | | *Rather unstable |
| trans-1,2-Diphenylethene (trans-Stilbene) | | 124 | | | | | | Unsaturated; dibromide, 237, formed on warming with bromine. Styphnic acid deriv., 142 (p. 38) |
| Pyrene | | 150 | | | | | 227 | Styphnic acid deriv., 191 |
| Anthracene | | 217 | | | | | 138 | Maleic anhydride adduct, 263 (p. 38). $CrO_3$-ethanoic acid → quinone, 286. Styphnic acid deriv., 180 (p. 38) |

### Table 26. Ketones (C, H and O)

| | B.p. | M.p. | 2,4-Di-nitro-phenyl-hydra-zone (p. 39) | Semi-carba-zone (p. 39) | 4-Nitro phenyl-hydra-zone (p. 39) | Notes |
|---|---|---|---|---|---|---|
| Propanone (Acetone) | 56 | | 126 | 187 | 148 | Monobenzylidene deriv., 42. Dibenzylidene deriv., 112 (p. 39) |
| Butan-2-one (Methyl ethyl ketone) | 80 | | 116 | 146 | 128 | |
| But-3-en-2-one (Methyl vinyl ketone) | 80 | | | 141 | | Unsaturated |
| Butane-2,3-dione (Diacetyl) | 88 | | 315 | 235 mono 278 di | 230 mono >310 di | Monobenzylidene deriv., 53 (p. 39) |
| 2-Methylbutan-3-one (Isopropyl methyl ketone) | 94 | | 124 (118) | 113 | 108 | Benzylidene deriv., 117 (p. 39) |
| Pentan-3-one (Diethyl ketone) | 102 | | 156 | 139 | 144 | Monobenzylidene deriv., 31. Dibenzylidene deriv., 127 |
| Pentan-2-one (Methyl propyl ketone) | 102 | | 143 | 110 | 117 | |
| 3,3-Dimethylbutan-2-one (Pinacolone) | 106 | | 125 | 158 | 139 | Benzylidene deriv., 41 |
| 3-Benzoylpropanoic acid | 116 | | 191 | 181 | | See Table 16 |
| 4-Methylpentan-2-one (Isobutyl methyl ketone) | 117 | | 95 | 130 | 79 | |
| 3-Methylpentan-2-one (s-Butyl methyl ketone) | 118 | | 71 | 94 | | |
| 2,4-Dimethylpentan-3-one (Di-isopropyl ketone) | 124 | | 96 (107) | 159* | | *Varies with the rate of heating |

**Table 26.** Ketones (C, H and O) (*cont.*)

| | B.p. | M.p. | 2,4-Di-nitro-phenyl-hydra-zone (p. 39) | Semi-carba-zone (p. 39) | 4-Nitro-phenyl-hydra-zone (p. 39) | Notes |
|---|---|---|---|---|---|---|
| Hexan-2-one | 128 | | 106 | 122 | 88 | |
| (Butyl methyl ketone) | | | | | | |
| 4-Methylpent-3-en-2-one | 130 | | 203 | 164 | 133* | *Prepared without heating. |
| (Mesityl oxide) | | | | (133) | | Unsaturated |
| Cyclopentanone | 131 | | 146 | 206 | 154 | Benzylidene deriv., 190 (p. 39) |
| Methyl 2-oxopropanoate | 136 | | 187 | 208 | | See Table 19 |
| (Methyl pyruvate) | | | | | | |
| 2-Methylhexan-4-one | 136 | | 75 | 152 | | |
| (Ethyl isobutyl ketone) | | | | | | |
| 2-Methylhexan-3-one | 136 | | 97 | 119 | | |
| (Isopropyl propyl ketone) | | | | | | |
| Pentane-2,4-dione | 139 | | 209 di 122* | 107* | | *Pyrazole deriv., Oxime, 149 |
| (Acetylacetone) | | | | | | (p. 39). See Table 18 |
| 2-Methylcyclopentanone | 139 | | | 184 | | |
| 5-Methylhexan-2-one | 144 | | 95 | 143 | | |
| (Isopentyl methyl ketone) | | | | | | |
| Heptan-4-one | 144 | | 75 | 133 | | |
| (Dipropyl ketone) | | | | | | |
| 3-Hydroxybutan-2-one | 145 | | 315 | 185 | | |
| (Acetoin) | | | | | | |
| Hydroxypropanone | 146 | | 129 | 196 | 173 | |
| (Hydroxyacetone, Acetol) | | | | | | |
| Heptan-2-one | 151 | | 89 | 123 | 73 | |
| (Methyl pentyl ketone) | | | | | | |
| 2-Methylheptan-4-one | 155 | | | 123 | | |
| (Isobutyl propyl ketone) | | | | | | |
| Cyclohexanone | 155 | | 162 | 166 | 146 | Benzylidene deriv., 118 (p. 39) |
| Ethyl 2-oxopropanoate | 155 | | 155 | 206d | | See Table 19 |
| (Ethyl pyruvate) | | | | | | |
| 3,5-Dimethylheptan-4-one | 162 | | | 84 | | |
| (Di-s-butyl ketone) | | | | | | |
| 2-Methylcyclohexanone | 163 | | 136 | 196 | 132 | |
| 4-Hydroxy-4-methylpentan-2-one | 165 | | 203 | | 209 | |
| (Diacetone alcohol) | | | | | | |
| 2-Oxopropanoic acid | 165d | | 218 | 222 | 220 | See Table 19 |
| (Pyruvic acid) | | | | | | |
| 2,6-Dimethylheptan-4-one | 168 | | 92 (66) | 122 | | |
| (Di-isobutyl ketone) | | | | | | |
| 3-Methylcyclohexanone | 168 | | 155 | 191 (180) | 119 | Benzylidene deriv., 122 (p. 39) |
| 4-Methylcyclohexanone | 169 | | 134 | 198 | 128 | Benzylidene deriv., 99 (p. 39) |
| Methyl-3-oxobutanoate | 170 | | 119 | 152 | | Gives red colour with aq. $FeCl_3$. |
| (Methyl acetoacetate) | | | | | | See Tables 18 and 19 |
| 2-Methylheptan-6-one | 171 | | 77 | 156 | | |
| (Isohexyl methyl ketone) | | | | | | |
| Octan-2-one | 173 | | 58 | 122 | 93 | |
| (Hexyl methyl ketone) | | | | | | |
| Cyclohexyl methyl ketone | 180 | | 140 | 177 | 154 | |
| Ethyl 3-oxobutanoate | 181d | | 96 | 133 | 218* | Gives red colour with aq. $FeCl_3$. |
| (Ethyl acetoacetate) | | | | | | See Tables 18 and 19. |
| | | | | | | *Pyrazole derivative |

83

**Table 26.** Ketones (C, H and O) (*cont.*)

| | B.p. | M.p. | 2,4-Di-nitro-phenyl-hydra-zone (p. 39) | Semi-carba-zone (p. 39) | 4-Nitro-phenyl-hydra-zone (p. 39) | Notes |
|---|---|---|---|---|---|---|
| Cycloheptanone | 181 | | 148 | 162 | 137 | Benzylidene deriv., 108 (p. 39) |
| Nonan-5-one | 187 | | 41 | 90 | | |
| (Dibutyl ketone) | | | | | | |
| Hexane-2,5-dione | 190 | | 256 | 185 | 115 | |
| (Acetonylacetone) | | | | mono | | |
| | | | | 220 | | |
| | | | | di | | |
| Nonan-2-one | 194 | | 56 | 120 | | |
| Fenchone | 194 | | 140 | 183 | | Unsaturated |
| Methyl 4-oxopentanoate | 196 | | 141 | 144 | 136 | |
| (Methyl laevulinate) | | | | | | |
| Cyclo-octanone | 196 | | 163 | 167 | | |
| 2,6-Dimethylhepta-2,5- | | | | | | |
| dien-4-one | 198 | 28 | 112 | 186 | | |
| (Phorone) | | | | | | |
| α-Thujone | 200 | | 116 | 186 | | |
| β-Thujone | 202 | | 114 | 174 | | |
| Acetophenone | 202 | 20 | 249 | 198 | 184 | Benzylidene deriv., 58 (p. 39) |
| | | | (240) | | | |
| Ethyl 4-oxopentanoate | 206 | | 101 | 150 | 157 | |
| (Ethyl laevulinate) | | | | | | |
| (−)-Menthone | 207 | | 146 | 184 | | |
| Decan-2-one | 209 | 14 | 74 | 124 | | |
| Decan-3-one | 211 | | | 101 | | |
| 3,5,5-Trimethylcyclohex-2- | | | | | | |
| en-1-one | 214 | | 130 | 191 | | |
| (Isophorone) | | | | | | |
| 1-Phenylpropan-2-one | 216 | 27 | 156 | 190 | 145 | |
| (Benzyl methyl ketone) | | | | (196) | | |
| 2-Methylacetophenone | 216 | | 159 | 206 | | |
| (Methyl o-tolyl ketone) | | | | | | |
| Propiophenone | 218 | 18 | 191 | 174 | 147 | |
| 2-Hydroxyacetophenone | 218 | 28 | 213 | 210 | | Oxime, 117 (p. 39). See Table 30 |
| (o-Acetylphenol) | | | | | | |
| 3-Methylacetophenone | 220 | | 207 | 200 | | |
| (Methyl m-tolyl ketone) | | | | | | |
| Isopropyl phenyl ketone | 222 | | 163 | 181 | | |
| (Isobutyrophenone) | | | | | | |
| Butyl phenyl ketone | 230 | | 190 | 190 | | |
| (Butyrophenone) | | | | | | |
| 4-Methylacetophenone | 223 | | 260 | 205 | | |
| (Methyl p-tolyl ketone) | | | | | | |
| (+)-Pulegone | 224 | | 147 | 174 | | Unsaturated |
| (+)-Carvone | 225 | | 190 | 162 | 174 | Unsaturated |
| | | | | (142) | | |
| 1-Phenylbutan-2-one | 226 | | 140 | 135 | | |
| (Benzyl ethyl ketone) | | | | (146) | | |
| Undecan-2-one | 228 | | 63 | 122 | 90 | |
| 1-Phenylbutan-3-one | 235 | | 128 | 142 | | |
| 3-Methoxyacetophenone | 240 | | 189 | 196 | | |
| Butyl phenyl ketone | 242 | | 166 | 166 | 162 | |
| (Valerophenone) | | | | | | |
| 2-Methoxyacetophenone | 245 | | | 183 | | Oxime, 83 |

# Table 26. Ketones (C, H and O) (*cont.*)

| | B.p. | M.p. | 2,4-Di-nitro-phenyl-hydra-zone (p. 39) | Semi-carba-zone (p. 39) | 4-Nitro-phenyl-hydra-zone (p. 39) | Notes |
|---|---|---|---|---|---|---|
| 4-Oxopentanoic acid (Laevulinic acid) | 245d | 33 | 206 | 187d | 175 | Monobenzylidene deriv., 123 (p. 39). See Table 16 |
| Diethyl 3-oxopentanedioate (Ethyl acetonedicarboxylate) | 250 | | 86 | 94 | | Red colour with aq. FeCl$_3$. See also Table 19 |
| α-Tetralone | 257 | | 257 | 226 | 231 | Benzylidene deriv., 105 (p. 39) |
| 4-Methoxyacetophenone | 258 | 38 | 227 | 197 | 195 | |
| Methyl 1-naphthyl ketone | 298 | 34 | 255 | 235 | | Benzylidene deriv., 126 (p. 39) |
| 1,2-Diphenylpropanone (Dibenzyl ketone) | 330 | 34 | 100 | 146 | | Monobenzylidene deriv., 162 (p. 39) |
| Benzylidenepropanone | 262 | 41 | 227 | 186 | 166 | Unsaturated. Benzylidene deriv., 112 (p. 39) |
| Indanone | | 42 | 258 | 233 | 235 | Benzylidene deriv., 113 (p. 39) |
| Diphenyl ketone (Benzophenone) | | 48 | 238 | 165 | 154 | |
| Methyl 2-naphthyl ketone | | 53 | 262d | 236 | | |
| Phenyl 4-tolyl ketone | | 54 (60) | 200 | 122 | | |
| Benzylideneacetophenone (Chalcone) | | 58 | 248 (208) | 168 (180) | | Unsaturated |
| Benzyl phenyl ketone (Deoxybenzoin) | | 60 | 204 | 148 | 163 | Benzylidene deriv., 102 (p. 39) |
| 1-Phenylbutane-1,3-dione (Benzoylacetone) | | 61 | 151 | | 101 | Monobenzylidene deriv., 99. See Table 18 |
| 4-Methoxybenzophenone | | 62 | 180 228* | | 199 | *From trichloromethane |
| Fluorenone | | 83 | 284 | 234 | 269 | Yellow |
| Di-4-tolyl ketone | | 93 | 229 | 140 | | Oxime, 163 (p. 39) |
| Benzil | | 95 | 189 mono 314 di | 244d di 182 mono | 290 | Yellow |
| 3-Hydroxyacetophenone (*m*-Acetylphenol) | | 96 | 256 | 195 | | See Table 30 |
| 4-Hydroxyacetophenone (*p*-Acetylphenol) | | 110 | 261 (225) | 199 | | See Table 30 |
| Dibenzylidenepropanone | | 112 | 180 | 190 | 172 | Unsaturated |
| 3-Benzoylpropanoic acid | | 116 | 191 | 181 | | See Table 16 |
| 4-Phenylacetophenone | | 121 | 241 | | | Oxime, 187 (p. 39) |
| 2-Benzoylbenzoic acid | | 128 | | | | Oxime, 118; see Table 16 |
| 4-Hydroxy-3-methoxybenzylidene-propanone (Vanillideneacetone) | | 130 | 230 | | | Unsaturated |
| Benzoylphenylmethanol (Benzoin) | | 133 | 245 | 206d | | Ethanoyl deriv., 83. See Table 2 |
| 4-Hydroxybenzophenone | | 135 | 242 | 194 | | See Table 30 |
| Furoin | | 136 | 216 | | | Oxime, 161 (p. 39) |
| 2,4-Dihydroxyacetophenone (Resacetophenone) | | 147 | 218 (244) | 216 | | |
| 4-Hydroxypropiophenone | | 148 | 229 240* | | | *From trichloromethane |
| Furil | | 162 | 215 | | 199d | |
| 2,3,4-Trihydroxyacetophenone (Gallacetophenone) | | 173 | 199 | 225 | | Oxime, 163; triethanoate, 85; see Table 30 |
| (+)-Camphor | | 179 | 177 | 237 | 217 | Benzylidene deriv., 98 (p. 39) |

**Table 27.** Ketones (C, H, O and halogen or N)

| | B.p. | M.p. | 2,4-Di-nitro-phenyl-hydra-zone (p. 39) | Semi-carba-zone (p. 39) | 4-Nitro-phenyl-hydra-zone (p. 39) | Notes |
|---|---|---|---|---|---|---|
| Chloropropanone | 119 | | 125 | 164* | 83 | *Variable |
| 1,1-Dichloropropanone | 120 | | | 163 | | |
| 4-Fluoroacetophenone | 196 | | 235 | 219 | | Oxime, 80 (p. 39) |
| 3-Chloroacetophenone | 228 | | | 232 | 176 | Oxime, 88 (p. 39) |
| 4-Chloroacetophenone | 232 | | 236 | 201 | 239 | |
| 2-Aminoacetophenone | 250d | 20 | | 290d | | Oxime, 109 (p. 39). See Table 9 |
| 2-Nitroacetophenone | 178/ 32 mm | 27 | 154 | 210d | | Oxime, 117 (p. 39) |
| 4-Chloropropiophenone | 134/ 31 mm | 36 | 222 | 176 | | |
| Phenacyl bromide | | 50 | 220 | 146 | | |
| 4-Bromoacetophenone | 255 | 51 | 230 | 208 | 248 | |
| Phenacyl chloride | | 59 | 212 | 156 | | |
| 4-Chlorobenzophenone | | 78 | 185 | | | Oxime, 163 (p. 39) |
| 3-Nitroacetophenone | | 80 | 232 | 259 | | |
| 3-Aminoacetophenone | | 99 | | 196 | | See Table 9 |
| 4-Aminoacetophenone | | 106 | 259 | 250 | | See Table 9 |
| 4-Bromophenacyl bromide | | 108 | 218 | | | Oxime, 115 (p. 39). Benzoic acid ester, 119 |
| 4-Phenylphenacyl bromide | | 126 | 228 | | | Benzoic acid ester, 167 |
| 4,4'-Di-(dimethylamino)benzo-phenone (Michler's ketone) | | 174 | 273 | | | Oxime, 233 (p. 39). See Table 12 |

**Table 28.** Nitriles (some nitriles are also listed in Table 16)

| | B.p. | M.p. | Car-boxylic acid (p. 40) | 4-Bromo-phenacyl ester of acid (p. 32) | Amide (p. 39) | Notes |
|---|---|---|---|---|---|---|
| Propenenitrile (Acrylonitrile) | 78 | | | | | Unsaturated. 2-Naphthol adduct, 142 |
| Ethanenitrile (Acetonitrile) | 82 | | | 85 | | |
| Propanenitrile (Propionitrile) | 97 | | | 61 | | |
| 2-Methylpropanenitrile (Isobutyronitrile) | 108 | | | 77 | | |
| Butanenitrile (Butyronitrile) | 118 | | | 63 | | |
| But-3-enenitrile (Allyl cyanide) | 118 | | | 60 | | Unsaturated |
| Chloroethanenitrile (Chloracetonitrile) | 127 | | | 120 | | |
| 3-Methylbutanenitrile (Isovaleronitrile) | 129 | | | 68 | | |
| Pentanenitrile (Valeronitrile) | 140 | | | 75 | | |

**Table 28.** Nitriles (some nitriles are also listed in Table 16) (*cont.*)

| | B.p. | M.p. | Carboxylic acid (p. 40) | 4-Bromo-phenacyl ester of acid (p. 32) | Amide (p. 39) | Notes |
|---|---|---|---|---|---|---|
| 4-Methylpentanenitrile (Isocapronitrile) | 155 | | | 77 | | |
| Hexanenitrile (Capronitrile) | 163 | | | 72 | | |
| 2-Hydroxy-2-phenylethane-nitrile (Mandelonitrile) | 170d | 21 | 118 | 113 | | |
| Benzonitrile | 191 | | 122 | 119 | 128 | Smell of bitter almonds |
| 2-Toluonitrile | 205 | | 104 | 57 | 142 | |
| 3-Toluonitrile | 212 | | 111 | 108 | 96 | |
| 4-Toluonitrile | 218 | 29 | 180 | 153 | 160 | |
| Propanedinitrile (Malononitrile) | 219 | | 133d | | 170 | |
| Phenylethanenitrile (Benzyl cyanide) | 232 | | 76 | 89 | 157 | |
| Hexanedinitrile (Adiponitrile) | 295 | | 153 | 155 | 220 | |
| 1-Naphthonitrile | 299 | 35 | 161 | 135 | 202 | |
| 3-Chlorobenzonitrile | | 41 | 158 | 117 | 134 | |
| 2-Chlorobenzonitrile | | 43 | 142 | 106 | 142 | |
| Butanedinitrile (Succinonitrile) | | 54 | 185 | 211 | 260 | |
| 4-Chlorobenzonitrile | | 92 | 241 | 126 | 179 | |
| Benzene-1,2-dinitrile (Phthalonitrile) | | 141 | 200d | 153 | 220 | |
| 4-Nitrobenzonitrile | | 148 | 240 | 134 | 201 | |

**Table 29.** Nitro-, halogenonitro-compounds and nitro-ethers

| | B.p. | M.p. | Nitro deriv. (p. 34) position | M.p. | meth-od | Colour with aq. NaOH | Notes |
|---|---|---|---|---|---|---|---|
| Nitromethane | 101 | | | | | | Acid to litmus; benzylidene deriv., 58 (p. 40) |
| Nitroethane | 114 | | | | | | Benzylidene deriv., 64 (p. 40) |
| 2-Nitropropane | 120 | | | | | | Reduction with Sn + HCl → 2-aminopropane (p. 40) |
| 1-Nitropropane | 132 | | | | | | Immiscible with water. Sn + HCl → propylamine (p. 40) |
| Nitrobenzene | 211 | | 1,3 | 90 | ii | | Pale yellow; odour of bitter almonds. Sn + HCl → aniline (p. 40) |
| 2-Nitrotoluene | 222 | | 2,4 | 70 | ii | | Pale yellow; odour of bitter almonds. Sn + HCl → 2-toluidine (p. 40) |
| 1,3-Dimethyl-2-nitro-benzene (2-Nitro-*m*-xylene) | 226 | 13 | 2,4,6 | 182 | ii | | |
| Phenylnitromethane | 226d | | | | | | Yellow; benzylidene deriv., 75 |

**Table 29.** Nitro-, halogenonitro-compounds and nitro-ethers (*cont.*)

| | B.p. | M.p. | Nitro deriv. (p. 34) position | M.p. | method | Colour with aq. NaOH | Notes |
|---|---|---|---|---|---|---|---|
| 3-Nitrotoluene | 233 | 16 | | | | | Pale yellow; Sn + HCl → 3-toluidine (p. 40) Boiling aq. $K_2Cr_2O_7-H_2SO_4$ → acid, 140 (p. 40) |
| 6-Chloro-2-nitrotoluene | 238 | 37 | | | | | Pale yellow. $K_2Cr_2O_7-H_2SO_4$ → acid, 161 (p. 40) |
| 1,4-Dimethyl-2-nitrobenzene (2-Nitro-*p*-xylene) | 240 | | 2,3,5 | 139 | ii | | |
| 4-Ethylnitrobenzene | 241 | | | | | | Sn + HCl → 4-ethylaniline (p. 40) |
| 1,3-Dimethyl-4-nitrobenzene (4-Nitro-*m*-xylene) | 244 | 2 | 2,4,6 | 182 | ii | | |
| 2-Chloronitrobenzene | 246 | 32 | 2,4 | 52 | ii | | Pale yellow |
| 1,2-Dimethyl-3-nitrobenzene (3-Nitro-*o*-xylene) | 250 | 15 | 3,4 | 82 | ii | | Pale yellow |
| 1,2-Dimethyl-4-nitrobenzene (4-Nitro-*o*-xylene) | 258 | 29 | 3,4 | 82 | ii | | |
| 2-Methoxynitrobenzene (*o*-Nitroanisole) | 265 | 9 | 2,4 / 2,4,6 | 88* / 68 | i / ii | | *Nitration at 0° |
| 2-Ethoxynitrobenzene (*o*-Nitrophenetole) | 267 | | 2,4 / 2,4,6 | 86* / 78 | i / ii | | *Nitration at 0° |
| 2-Nitrobiphenyl | 320 | 37 | 2,4' | 93 | ii | | |
| 2-Bromonitrobenzene | 259 | 41 | 2,4 | 72 | ii | | Pale yellow |
| 2-Nitro-1,3,5-trimethyl-benzene (Nitromesitylene) | 255 | 44 | 2,4 / 2,4,6 | 86 / 235 | iv / ii | | |
| 3-Chloronitrobenzene | 236 | 44 | 3,4 | 36 | ii | | Pale yellow; Sn + HCl → 3-chloroaniline (p. 40) |
| 4-Nitrotoluene | 234 | 52 | 2,4 | 70 | ii | | Pale yellow; odour like nitro-benzene. $K_2Cr_2O_7$-dil. $H_2SO_4$ → acid, 241 (p. 40) |
| 1-Chloro-2,4-dinitrobenzene | | 52 | 2,4,6 | 183 | ii | Red → lilac | Reactive chlorine atom; boiling 2N NaOH → 2,4-dinitrophenol, 114 Hydrazine → 2,4-dinitrophenyl-hydrazine, 199 |
| 4-Methoxynitrobenzene (*p*-Nitroanisole) | | 53 | 2,4 | 87 | i | | Boiling conc. NaOH → 4-nitro-phenol, 114 |
| 3-Bromonitrobenzene | | 56 | 3,4 | 59 | ii | | Pale yellow |
| 1,4-Dichloro-2-nitrobenzene | | 56 | 2,6 | 104 | ii | | Pale yellow. KOH in boiling aq. methanol → 4-chloro-2-nitro-methoxybenzene, 98 |
| 1-Nitro-2-phenylethene (β-Nitrostyrene) | | 58 | | | | | Yellow. Sn + HCl → 2-phenyl-ethylamine (p. 40) |
| 4-Ethoxynitrobenzene | | 60 | 2,4 / 2,4,6 | 86 / 78 | i / ii | | Boiling with 40% HBr → 4-nitro-phenol, 114 |
| 1-Nitronaphthalene | | 60 | 1,3,8 | 218 | ii | | Yellow. Picrate, 71. $CrO_3$-ethanoic acid → 3-nitrobenzene-1,2-di-carboxylic acid, 218 |
| 2,6-Dinitrotoluene | | 66 | | | | Violet | Boiling dil. $HNO_3$ → acid, 202 |

**Table 29.** Nitro-, halogenonitro-compounds and nitro-ethers (*cont.*)

| | B.p. | M.p. | Nitro deriv. (p. 34) position | Nitro deriv. (p. 34) M.p. | method | Colour with aq. NaOH | Notes |
|---|---|---|---|---|---|---|---|
| 1-Methoxy-2,4,6-trinitrobenzene (2,4,6-Trinitroanisole) | | 68 | | | | Purple | Yellow. Boiling dil. NaOH → picric acid, 122. NH$_3$ in ethanol → 2,4,6-trinitroaniline, 188 |
| 2,4-Dinitrotoluene | | 70 | 2,4,6* | 82 | ii | Blue | CrO$_3$-conc. H$_2$SO$_4$ → acid, 183 (p. 40) *Explosive; not recommended |
| 1-Bromo-2,4-dinitrobenzene | | 72 | | | | Red | Pale yellow. Reactive bromine atom; boiling 2M NaOH → 2,4-dinitrophenol, 114. Hydrazine → 2,4-dinitrophenylhydrazine, 199 |
| 1,3-Dimethyl-5-nitrobenzene (5-Nitro-*m*-xylene) | | 75 | 4,5,6 | 125 | ii | | |
| 1-Ethoxy-2,4,6-trinitrobenzene (2,4,6-Trinitrophenetole) | | 78 | | | | Red | Yellow. Boiling dil. NaOH → picric acid, 122 |
| 2,4,6-Trinitrotoluene (T.N.T.) | | 82 | | | | Red | Explosive. Addition product with naphthalene, 97 |
| 4-Chloronitrobenzene | | 83 | 2,4 | 52 | ii | | Pale yellow. Reactive chlorine atom; boiling aq. KOH → 4-nitrophenol, 114 |
| 1-Chloro-2,4,6-trinitrobenzene (Picryl chloride) | | 83 | | | | Red | Yellow. Reactive chlorine atom; warm aq. KOH → picric acid, 122. Naphthalene adduct, 150 |
| 1,4-Dibromo-2-nitrobenzene | | 84 | | | | | Pale yellow. Reactive bromine atom; boiling aq. methanolic KOH → 4-bromo-2-nitro-methoxybenzene, 86 |
| 1-Methoxy-2,4-dinitrobenzene (2,4-Dinitroanisole) | | 87 | 2,4,6 | 66 | ii | Purple | Pale yellow. Boiling aq. KOH → 2,4-dinitrophenol, 114 |
| 1,3-Dinitrobenzene | | 90 | | | | Purple | Pale yellow. Hot ethanolic NH$_4$SH → 3-nitroaniline, 114 (p. 41) |
| 1,3-Dimethyl-4,6-dinitrobenzene (4,6-Dinitro-*m*-xylene) | | 93 | 2,4,6 | 182 | ii | Violet | Pale yellow. Hot ethanolic NH$_4$SH → 2,4-dimethyl-5-nitroaniline, 123 (p. 41) |
| 4-Nitrobiphenyl | | 113 | 4,4 | 233 | ii | | Carcinogenic |
| 1,2-Dinitrobenzene | | 118 | | | | None | Hot ethanolic NH$_4$SH → 2-nitroaniline, 71; hot aq. NaOH → 2-nitrophenol, 45 |
| 1,3,5-Trinitrobenzene | | 122 | | | | Red | Pale yellow. Naphthalene adduct (in ethanol), 152 |
| 4-Bromonitrobenzene | | 126 | 2,4 | 72 | ii | | Pale yellow. Reactive bromine atom; boiling aq. KOH → 4-nitrophenol, 114 |
| 1,4-Dinitrobenzene | | 172 | | | | Green-yellow | Naphthalene adduct (in ethanol), 118. Boiling ethanolic NH$_4$SH → 4-nitroaniline, 147 (p. 41) |

**Table 30.** Phenols (C, H and O)

| | B.p. | M.p. | FeCl₃ colour Aq. | FeCl₃ colour MeOH | Benzoate (p. 41) | Aryloxy-ethanoic acid (p. 41) | Toluene-4-sulphonate (p. 41) | 3,5-Dinitro-benzoate (p. 41) | Notes |
|---|---|---|---|---|---|---|---|---|---|
| 2-Methylphenol (o-Cresol) | 190 | 31 | B→G | G | Oil | 152 | 55 | 138 | 4-Nitrobenzoate, 128 (p. 41) |
| 2-Hydroxybenzaldehyde (Salicylaldehyde) | 196 | | V | V | | 132 | 63 | | See Table 4 |
| 3-Methylphenol (m-cresol) | 202 | 12 | B→G | G | 55 | 103 | 51 | 165 | |
| 4-Methylphenol (p-cresol) | 202 | 35 | B | YG | 71 | 136 | 70 | 188 | |
| 2-Methoxyphenol (Guaiacol) | 205 | 30 | R | G | 57 | 116 | 82 | 142 | |
| 2-Ethylphenol | 207 | | B | G | 38 | 141 | | 108 | |
| 2,4-Dimethylphenol (1,3-Xylen-4-ol) | 211 | 27 | B | GBt | 38 | 142 | | 164 | |
| 2-Hydroxyacetophenone | 218 | 28 | VR | VR | 87 | | | | See Table 26 |
| 3-Ethylphenol | 216 | | V | G | 52 | 75 | | | |
| Methyl 2-hydroxybenzoate (Methyl salicylate) | 224 | | V | | 92 | | | | 4-Nitrobenzoate, 128 (p. 41) See Table 19 |
| Ethyl 2-hydroxybenzoate (Ethyl salicylate) | 233 | | RV | V | 79 (87) | | | | 4-Nitrobenzoate, 107 (p. 41) See Table 19 |
| 4-(2-Methylpropyl)phenol | 236 | | | | | 125 | | | |
| 2-Methyl-5-isopropylphenol (Carvacrol) | 237 | | | Gt | | 151 | | 80 | |
| 3-Methoxyphenol | 243 | | V | V | | 116 | | | |
| 4-Butylphenol | 248 | 22 | | | 27 | 81 | | | 4-Nitrobenzoate, 68 (p. 41) |
| 4-(Prop-2-enyl)-2-methoxyphenol (Eugenol) | 253 | | YG | B | 70 | 100* | 85 | 131 | *Hydrate, 81. Unsaturated |
| Butyl 2-hydroxybenzoate (Butyl salicylate) | 259 (270) | | V | V | | | | | Nitration (method ii, p. 34) → 3,5-dinitro deriv., 61. See Table 19 |
| 2-Methylpropyl 2-hydroxybenzoate (Isobutyl salicylate) | 261 | | V | V | | | | | Nitration (method ii, p. 34) → 3,5-dinitro deriv., 72. See Table 19 |
| 2-Methoxy-4-(prop-1-enyl)-phenol (Isoeugenol) | 267 | | Gt | Gt | 106 | 94 | | 158 | Unsaturated. Dibromide, 94 |
| Phenol | 182 | 42 | V | G | 69 | 99 | 96 | 146 | 4-Nitrobenzoate, 111 (p. 41) See Table 19 |
| Phenyl 2-hydroxybenzoate (Salol) | | 42 | | VR | 81 | | | | |
| 4-Ethylphenol | 219 | 47 | B | | 60 | 97 | | 132 | |
| 2,6-Dimethylphenol | 203 | 49 | Y | Y | 39 | 140 | | 159 | |
| 3-Methyl-6-isopropylphenol (Thymol) | 233 | 50 | – | RBr | 33 | 148 | 71 | 103 | |

| Compound | | m.p. | Colour | | | | | | |
|---|---|---|---|---|---|---|---|---|---|
| 4-Methoxyphenol | 243 | 54 | Vt | G | 87 | 111 | | | Nitration (method ii, p. 34) → 4,6-dinitro deriv., 106 |
| 2-Cyclohexylphenol | | 56 | | | 40 | | 65 | | |
| 2-Phenylphenol | | 57 (67) | | | 76 | 107 | | | Heating at 100° → anhyd. form, 107 |
| 3,5-Dihydroxytoluene, hydrate (Orcinol) | | 58 | BV | — | 88 | 217 | | 190 | |
| 3,4-Dimethylphenol | | 62 | B | Y | 59 | 163 | | 182 | |
| 3,5-Dimethylphenol | | 68 | | GB | 24 | 111* | 83 | 195 | *Anhydrous; hydrate, 84 |
| 2,4,5-Trimethylphenol | | 71 | — | | 63 | 132 | | 179 | |
| 2,5-Dimethylphenol | | 75 | | YG | 61 | 118 | | 137 | |
| 2,3-Dimethylphenol | | 75 | B | | 78 | 187 | | | |
| 4-Hydroxy-3-methoxybenzaldehyde (Vanillin) | | 80 | BV | G | 82 | 188 | 115 | | 4-Nitrobenzoate, 104 (p. 41) |
| 4-(1,1,3,3-Tetramethylbutyl)phenol (p-t-Octylphenol) | | 84 | | | 87 | 108 | | | See Table 4 |
| 4-Benzylphenol (p-Hydroxydiphenyl-methane) | | 84 | | G | | | | | |
| 2-Hydroxybenzyl alcohol (Saligenin) | | 86 | | G | 51* | 120 | 75 | | *Dibenzoate, 85 |
| 4-(2-Methylbut-2-yl)phenol | | 92 | Pk* | Gt | 61 | | 55 | | |
| 1-Naphthol | | 94 | | Br | 56 | 192 | 89 | 217 | *Colour of precipitate See Table 26 |
| 3-Hydroxyacetophenone | | 96 | | | 52 | | | | |
| 4-(2-Methylprop-2-yl)phenol (p-t-Butylphenol) | | 99 | | G | 82 | 86 | 110 | | See Table 4 |
| 3-Hydroxybenzaldehyde | | 104 | V | — | 38 | | | | 4-Nitrobenzoate, 170 (p. 41) |
| 1,2-Dihydroxybenzene (Catechol) | | 105 | G | G | 84 di / 131 mono | 148 | | 152 | |
| 3,5-Dihydroxytoluene, anhyd. (Orcinol) | | 107 | V | — | 88 | 217 | | 190 | Ethanoate, 54. See Table 26 |
| 4-Hydroxyacetophenone | | 110 | V | BrR | 134 | 177 | 72 | 138 | |
| 1,3-Dihydroxybenzene (Resorcinol) | | 110 | V | G | 117 di / 135 mono | 195 | 80 di | 201 | |
| Ethyl 4-hydroxybenzoate | | 115 | V | — | 94 | | | | See Table 19 |
| 1,3,5-Trihydroxybenzene, dihydrate (Phloroglucinol) | | 117 | V | G | 174 | | | 162 | |
| 4-Hydroxybenzaldehyde | | 117 | | Y | 91 | 198 | | | Ethanoate, 104 |
| 2,6-Dihydroxytoluene (2-Methylresorcinol) | | 119 | VBr | Br | 106 | | | | See Table 4 |
| 2-Naphthol | | 123 | W* | Gt | 107 | 154 | 125 | 210 | *Opalescent |
| Methyl 4-hydroxybenzoate | | 131 | V | — | 135 | | | | Ethanoate, 85. See Table 19 |

**Table 30.** Phenols (C, H and O) (*cont.*)

| | B.p. | M.p. | FeCl₃ colour Aq. | FeCl₃ colour MeOH | Benzoate (p. 41) | Aryloxy-ethanoic acid (p. 41) | Toluene-4-sulphonate (p. 41) | 3,5-Dinitro-benzoate (p. 41) | Notes |
|---|---|---|---|---|---|---|---|---|---|
| 4-Cyclohexylphenol | | 132 | | | 118 | | | 168 | 4-Nitrobenzoate, 137 (p. 41) |
| 1,2,3-Trihydroxybenzene (Pyrogallol) | | 133 | R | G | 90* | 198 | | 205 | *Dibenzoate, 126; monobenzoate, 138 |
| 4-Hydroxybenzophenone | | 135 | V* | Br | 94 | | | | *Very weak. Ethanoate, 81. See Table 26 |
| 2,4-Dihydroxybenzaldehyde (Resorcylaldehyde) | | 135 | R | R | | | | | See Table 4 |
| 1,2,4-Trihydroxybenzene (Hydroxyquinol) | | 140 | R* | | 120 | | | | *In the presence of a trace of aq. NaOH. Ethanoate, 96 |
| 2,4-Dihydroxyacetophenone (Resacetophenone) | | 147 | R | | 80 di | | | | See Table 26 |
| 4-Hydroxypropiophenone | | 148 | | | | | | | See Table 26 |
| 2-Hydroxy-5-methylbenzoic acid (5-Methylsalicylic acid) | | 153 | VB | B | 155 | 185 | | | Ethanoate, 152. See Table 16 |
| 3,4-Dihydroxybenzaldehyde (Protocatechualdehyde) | | 154 | | | 96 | | | | See Table 4 |
| 3,5-Dihydroxybenzaldehyde | | 157 | | | | | | | See Table 4 |
| 2-Hydroxybenzoic acid (Salicylic acid) | | 158 | V | V | 132 | 191 | 154 | | Ethanoate, 135. See Table 16 |
| 2,3-Dihydroxynaphthalene | | 162 | B* | | 152 | | | | Ethanoate, 105. *Precipitate also present |
| 2-Hydroxy-3-methylbenzoic acid (3-Methylsalicylic acid) | | 163 | VB | V | | 204 | | | Ethanoate, 113. See Table 16 |
| 4-Phenylphenol | | 165 | | G | 149 | 190 | 179 | | |
| 1,4-Dihydroxybenzene (Quinol, hydroquinone) | | 171 | * | | 200 | 250 | 159 | 317 | *Oxidized to 1,4-benzoquinone |
| 2,3,4-Trihydroxyacetophenone (Gallacetophenone) | | 173 | Br | VBr | 118 | | | | Triethanoate, 85. See Table 26 |
| 1,4-Dihydroxynaphthalene | | 176 | | | 169 | | | | Diethanoate, 128 |
| 2,7-Dihydroxynaphthalene | | 186 | | | 139 | | | | Diethanoate, 136 |
| 1-Hydroxy-2-naphthoic acid | | 195 | B | BG | | 149 | 150 | | Ethanoate, 158. See Table 16 |
| 3-Hydroxybenzoic acid | | 200 | | | | 206 | 163 | | Ethanoate, 131. See Table 16 |
| 2,5-Dihydroxybenzoic acid (Gentisic acid) | | 200 | BV | B | | | | | Diethanoate, 118; 2-ethanoate, 172; 5-ethanoate, 131. See Table 16 |
| 4-Hydroxybenzoic acid | | 213 | O | O | 221 | 278 | 169 | | Ethanoate, 187. See Table 16 |
| 2,4-Dihydroxybenzoic acid | | 213d | | | 152 | | | | Diethanoate, 136. See Table 16 |

| Compound | B.p. | M.p. | FeCl₃ Aq. | FeCl₃ MeOH | Benzoate | Aryloxy-ethanoic acid | Toluene-4-sulphonate | 3,5-Dinitro-benzoate | Notes |
|---|---|---|---|---|---|---|---|---|---|
| 1,3,5-Trihydroxybenzene, anhydrous (Phloroglucinol) | | 217 | V | G | 174* tri | | | 162 | Triethanoate, 105. See Table 18; *Di, 126; mono, 196 |
| 3-Hydroxy-2-naphthoic acid | | 222 | | | 204 | | | | Ethanoate, 184. See Table 16 |
| 3,5-Dihydroxybenzoic acid | | 236d | | | 227 | | | | Diethanoate, 160. See Table 16 |
| 1,5-Dihydroxynaphthalene | | 265 | – | – | 235 | | | | Diethanoate, 160 |

Abbreviations for colours produced by ferric chloride: B, blue; Br, brown; G, green; O, orange; Pk, pink; R, red; t, transient; V, violet; W, white; Y, yellow; –, no colour.

Note. In the above ferric chloride tests, any deviation from the solvent stated will frequently invalidate the test.

Table 31. Phenols (C, H, O and halogen or N)

| | B.p. | M.p. | FeCl₃ colour | | Benzoate (p. 41) | Aryloxy-ethanoic acid (p. 41) | Toluene-4-sulphonate (p. 41) | 3,5-Dinitro-benzoate (p. 41) | Notes |
|---|---|---|---|---|---|---|---|---|---|
| | | | Aq. | MeOH | | | | | |
| 2-Chlorophenol | 175 | 7 | V | V | Oil | 144 | 74 | 143 | |
| 2-Bromophenol | 194 | 5 | V | V | | 142 | 78 | | |
| 3-Chloro-4-methylphenol | 196 | 33 | | G | 71 | 108 | | | |
| 3-Chlorophenol | 214 | 33 | | | 71 | 109 | | 156 | |
| 3-Bromophenol | 236 | 33 | | | 86 | 108 | 53 | | |
| 2-Bromo-4-chlorophenol | 123/10 mm | | | | 99 | 139 | | | |
| 2,4-Dibromophenol | 238 | 36 | V | YG | 98 | 153 | 120 | | |
| 3-Iodophenol | | 40 | | | 73 | 115 | 60 | 183 | |
| 4-Chlorophenol | 217 | 43 | BV | G | 86 | 156 | 71 | 186 | |
| 2,4-Dichlorophenol | 209 | 43 | VB | YG | 97 | 140 | 125 | | |
| 2-Iodophenol | | 43 | | | 34 | 135 | 80 | | |
| 2-Nitrophenol | 216 | 45 | – | Br | 59 | 156* | 83 | 155 | *Difficult to purify |
| 3-Methyl-4-nitrophenol | | 56 | | | 77 | | | | |
| 4-Chloro-2-isopropyl-5-methylphenol (p-Chlorothymol) | | 60 | Y | Y | 72 | | | 129 | |
| 4-Bromophenol | | 64 | V | YG | 102 | 157 | 94 | 191 | |
| 4-Chloro-3-methylphenol | | 66 (56) | B | YG | 86 | 178 | 98 | | |
| 2,4,6-Trichlorophenol | | 68 | – | – | 76 | 182 | 101 | 136 | Releases CO₂ from bicarbonate |
| 2,4,5-Trichlorophenol | | 68 | | | 93 | 157 | 115 | | Releases CO₂ from bicarbonate |
| 8-Hydroxyquinoline | | 75 | BG | V | 118 | | | | 4-Nitrobenzoate, 174 (p. 41). See Table 12 |

Table 31. Phenols (C, H, O and halogen or N) (cont.)

| | B.p. | M.p. | FeCl₃ colour Aq. | FeCl₃ colour MeOH | Benzoate (p. 41) | Aryloxy-ethanoic acid (p. 41) | Toluene-4-sulphonate (p. 41) | 3,5-Dinitro-benzoate (p. 41) | Notes |
|---|---|---|---|---|---|---|---|---|---|
| 4-(Dimethylamino)phenol | | 75 | | | | | 130 | | Ethanoate, 78. See Table 12 |
| 1-Bromo-2-naphthol | | 84 | − | − | 98 | | 121 | | |
| 3-(Dimethylamino)phenol | | 85 | | − | 94 | | | | See Table 12 |
| 4,6-Dinitro-2-methylphenol | | 86 | | | 133 | | 167 | | Ethanoate, 95 |
| 4-Iodophenol | | 94 | − | − | 119 | 156 | 99 | | |
| 2,4,6-Tribromophenol | | 95 | RV | − | 81 | 200 | 113 | 174 | Releases $CO_2$ from bicarbonate |
| 3-Nitrophenol | | 97 | V | V | 95 | 155 | 113 | 159 | |
| 5-Bromo-2-hydroxybenzaldehyde | | 106 | | | | | | | See Table 5 |
| 4-Nitrophenol | | 114 | R | Br | 142 | 184 | 97 | 188 | |
| 2,4-Dinitrophenol | | 114 | RBr | RBr | 132 | 148* | 121 | | *Difficult to purify |
| 4-Chloro-3,5-dimethylphenol | | 115 | BG | G | 68 | 141 | 103 | | |
| 3-Aminophenol | | 122 | Br | RBr | 153 | | 110* | 179 | *N-Mono deriv., 157, O-mono deriv., 96. See Table 9 |
| 2,4,6-Trinitrophenol (Picric acid) | | 122 | − | − | 163 | | | | Yellow; releases $CO_2$ from $NaHCO_3$; naphthalene adduct, 150 |
| 4-Nitrosophenol | | 125d | | | | | | | See Table 14 |
| 2-Hydroxybenzamide (Salicylamide) | | 139 | R | V | 143 | | 84 | 224 | O-Ethanoate, 138 |
| 2,6-Dibromo-4-nitrophenol | | 144 | Br | − | 220 N- | | 128 | | Ethanoate, 181 |
| 2-Amino-4,6-dinitrophenol (Picramic acid) | | 169 | | | | | 191 | | Red. See Table 9 |
| 2-Aminophenol | | 174 | RBr | R | 184 | | 146* (139) | | *N-Mono deriv.; O-mono deriv., 101 |
| 2,4,6-Trinitroresorcinol (Styphnic acid) | | 179 | | | | | | | Bright yellow. Naphthalene adduct, 168 |
| 4-Aminophenol | | 184d | V | V→Br | 234 | 196 | 168 di 252 N- | 178 | See Table 9 |
| Pentachlorophenol | | 190 | − | RBr | 164 | | 145 | | Ethanoate, 150. Releases $CO_2$ from bicarbonate |

Abbreviations for colours produced by ferric chloride: B, blue; Br, brown; G, green; O, orange; Pk, pink; R, red; t, transient; V, violet; W, white; Y, yellow; −, no colour.

Note. In the above ferric chloride tests, any deviation from the solvent stated will frequently invalidate the test.

94

**Table 32.** Quinones

| | Colour | M.p. | Oxime (p. 42) | Semicarbazone (p. 42) mono | di | Quinol (p. 42) | Notes |
|---|---|---|---|---|---|---|---|
| 5-Isopropyl-2-methyl-1,4-benzoquinone (Thymoquinone) | Yellow | 45 | 162 | 202d | 237 | 143 | |
| 2-Methyl-1,4-benzo-quinone (p-Toluoquinone) | Yellow | 69 | 134d mono 220d di | 178 | 240d | 124 | |
| 2-Methyl-1,4-naphtho-quinone | Yellow | 106 | 166 di 160 mono | 178 4- | 240d | 170 | |
| 1,4-Benzoquinone | Deep yellow | 115 | 144d mono 240d di | 166 | 243 | 170 | |
| 1,4-Naphthoquinone (α-Naphthoquinone) | Yellow | 125 | 198 mono 207d di | 247 | | 176 | |
| 1,2-Naphthoquinone (β-Naphthoquinone) | Red | 146d | 169 di 163 2- 109 1- | 184 | | 108* | *Anhydrous; hydrate, 60 |
| Quinhydrone | Dark green | 171 | | | | 170 | $K_2Cr_2O_7$ + dil. $H_2SO_4$ → 1,4-benzoquinone |
| 9,10-Phenanthraquinone | Orange | 206 | 158 mono 202d di | 220d | | 148 | |
| Acenaphthenequinone | Yellow | 261 | 230 mono | 192 | 271 | | |
| 9,10-Anthraquinone | Pale yellow | 286 | 224 mono | | | 180 | 4-Nitrophenylhydra-zone, 238 (p. 28) |

**Table 33.** Sulphonic acids and their derivatives

This table is arranged according to the boiling or melting point of the sulphonyl chloride because many of the acids do not have definite and reproducible values.

| Sulphonyl chloride | M.p. | Acid | Amide (p. 43) | Anilide (p. 43) | Benzyl thiouro-nium salt of acid (p. 43) | Xanthyl deriv. of amide (p. 43) | Notes |
|---|---|---|---|---|---|---|---|
| Methane- | * | † | 90 | 99 | | | *B.p. 161 †B.p. 167/10 mm |
| Ethane- | * | | 58 | 58 | 115 | | *B.p. 177 |
| Propane-2- | * | | 60 | 84 | | | *B.p. 61/9 mm |
| Propane-1- | * | † | 52 | | | | *B.p. 67/9 mm †B.p. 136/1 mm |
| 2-Toluene- | 10 | 57 | 156 | 136 | 170 | 183 | |
| 3-Toluene- | 12 | | 108 | 96 | | | |
| Benzene- | 14 | 66 | 155 | 110 | 148 | 206 | Benzoyl deriv. of amide, 147 |
| 2,5-Dimethylbenzene- | 25 | 48* | 148 | | 184 | 176 | *Dihydrate, 86 |
| 2,4-Dimethylbenzene- | 34 | 62 | 137 | 110 | 146 | 188 | |
| 2,5-Dichlorobenzene- | 38 | 93 | 181 | 160 | 170 | | Ethanoyl deriv. of amide, 214 |
| 3,4-Dimethylbenzene- | 51 | 64 | 144 | | 208 | 190 | |
| 4-Chlorobenzene- | 53 | 68 | 144 | 104 | 175 | | |
| Toluene-2,4-di- | 56 | | 190 | 189 | | | |
| 2,4,6-Trimethylbenzene- | 57 | 77 | 142 | 109 | | 203 | Ethanoyl deriv. of amide, 165 |
| Benzene-1,3-di- | 63 | | 229 | 144 | 214 | 170 | |
| Naphthalene-1- | 67 | 90 | 150 | 112 | 137 | | Benzoyl deriv. of amide, 194 |
| (+)-Camphor-10- | 67 | 193 | 132 | 120 | 210 | | |
| 4-Toluene- | 69 | 92 | 137* | 103 | 182 | 197 | *Hydrate, 105 |
| 4-Bromobenzene- | 75 | 103 | 166 | 119 | 170 | | |
| Naphthalene-2- | 76 | 91 | 217 | 132 | 191 | | Ethanoyl deriv. of amide, 145 |
| 2-Carboxybenzene- | 79 | 68 hyd. 134 anhyd. | 222* | 194 | 206 | 199 | *Saccharin (sulpho-imide). See Table 6 |
| Naphthalene-2,7-di- | 162 | | 242 | | 212 | | |
| Naphthalene-1,5-di- | 183 | 245 | 310 | 249 | 253 | | |
| Anthraquinone-2- | 197 | | 261 | 193 | 211 | | |
| Anthraquinone-1- | 214 | 218 | | 216 | 191 | | |
| 4-Hydroxybenzene- (Phenol-p-sulphonic) | | | 177 | 141 | 169 | | Warm Br$_2$-water → tri-bromophenol, 95 |
| 4-Aminobenzene- (Sulphanilic acid) | | >300d | 165* | | 182 | 208 | Dibenzoyl deriv. of amide, 268 *See Table 10 |

**Table 34.** Thioethers (sulphides)

|  | B.p. | M.p. | Sulphone (p. 44) | Notes |
|---|---|---|---|---|
| Dimethyl thioether | 38 | | 109 | |
| Diethyl thioether | 92 | | 73 | |
| Dipropyl thioether | 142 | | 29 | |
| Di-(2-methylpropyl) thioether | 172 | | 17* | *B.p. 265 |
| Dibutyl thioether | 182 | | 44 | |
| Methylthiophenyl | 188 | | 88 | |
| Ethylthiophenyl | 204 | | 41 | |
| Diphenyl thioether | 295 | | 128 | |
| Dibenzyl thioether | 150 | 49 | 150 | |
| Di-4-tolyl thioether | 158 | 57 | 158 | |
| Di-1-naphthyl thioether | | 110 | 187 | |
| Di-2-naphthyl thioether | | 151 | 177 | |

**Table 35.** Thiols and thiophenols

|  | B.p. | M.p. | 2,4-Dinitro-phenyl sulphide (p. 44) | H 3-nitro-phthaloyl deriv. (p. 44) | 3,5-Dinitro-benzoyl deriv. (p. 44) | Notes |
|---|---|---|---|---|---|---|
| Methanethiol | 6 | | 128 | | | |
| Ethanethiol | 36 | | 115 | 149 | 62 | |
| Propane-2-thiol | 58 | | 95 | 145 | 84 | |
| Propane-1-thiol | 68 | | 81 | 137 | 52 | |
| Prop-2-en-1-thiol | 90 | | 72 | | | |
| 2-Methylpropane-1-thiol | 88 | | 76 | 136 | 64 | |
| Butane-1-thiol | 98 | | 66 | 144 | 49 | |
| 3-Methylbutane-1-thiol | 117 | | 59 | 145 | 43 | |
| Pentane-1-thiol | 127 | | 80 | 132 | 40 | |
| Ethane-1,2-dithiol | 146 | | 248 | | | |
| Hexane-1-thiol | 111 | | 74 | | | |
| Cyclohexanethiol | 159 | | 148 | | | |
| 2-Hydroxyethanethiol | 160 | | 101 | | | |
| Benzenethiol (Thiophenol) | 169 | | 121 | 130 | 149 | |
| Propane-1,3-dithiol | 169 | | 194 | | | |
| Heptane-1-thiol | 176 | | 82 | 132 | 53 | |
| Toluene-α-thiol | 194 | | 130 | 137 | 120 | |
| Toluene-2-thiol (Thio-o-cresol) | 194 | 15 | 101 | | | |
| Toluene-3-thiol (Thio-m-cresol) | 195 | | 91 | | | |
| Octane-1-thiol | 199 | | 78 | | | |
| Naphthalene-1-thiol (α-Thionaphthol) | 209 | | 176 | | | |
| Decane-1-thiol | 114/13 mm | | 85 | | | |
| Toluene-4-thiol | 195 | 43 | 103 | | | |
| 4-Aminobenzenethiol (p-Aminothiophenol) | | 46 | | | | See Table 10 |
| 4-Chlorobenzenethiol | | 53 | 123 | | | |
| Naphthalene-2-thiol (β-Thionaphthol) | | 81 | 145 | | | |

# 6
# PHARMACEUTICAL COMPOUNDS

## INTRODUCTION

Organic compounds used in medicine are so numerous and varied in character that their identification may be difficult. However, the number in common use is much smaller and a selection of these is listed in Table P1—P6. In order to characterize these pharmaceutical compounds, the following procedure is recommended.

1. Establish the elemental composition of the compound by the Lassaigne procedure (p. 1).

2. Determine the melting point or boiling point (p. 3—4).

3. By correlating these two results, it may be possible to make a tentative identification of the compound by reference to the appropriate table (P1—P6).

4. Further confirmation of the identity of the compound should be obtained from the chemical tests and spectral data given in Tables I—XII (pp. 10—21).

5. Reference may now be made to the *British Pharmacopoeia* (or other recognized pharmacopoeia) for specific tests and spectral data which may be given for the compound.

6. Where appropriate, the compound should be converted into a crystalline derivative chosen from those given in Chapter 5.

*Notes:* (*a*) Some drugs exist as metallic salts of acids and others as acid salts of bases; these usually have poorly defined melting points. For such compounds, it is necessary to liberate the free acids or bases by acidification or basification, respectively. The free acid or base should be characterized in the usual way.

(*b*) For a more comprehensive list of compounds, see *Isolation and Identification of Drugs* by E. G. C. Clarke (The Pharmaceutical Press, London).

(*c*) Compounds which are also listed in the tables in Chapter 5 are indicated by the appropriate table number.

(*d*) Sugars are listed in Table 15 of Chapter 5.

## Table P1. Compounds containing C, H, (and O)

| | B.p. | | M.p. |
|---|---|---|---|
| Methylpentynol | 120 | Phenindione | 151 |
| Eugenol (Table 30) | 253 | Testosterone | 154 |
| | | Methyltestosterone | 164 |
| | | Hydrocortisone hydrogen succinate | 168 |
| | M.p. | Stilboestrol | 172 |
| Salol (Tables 9 and 30) | 42 | Oestradiol | 175 |
| Butylated hydroxyanisole | 62 | Camphor | 181 |
| Hexylresorcinol | 66 | Hexoestrol | 185 |
| Vanillin (Tables 4 and 30) | 80 | Ascorbic acid | 191d |
| Dimethisterone | 100d | Norethisterone | 206 |
| Acetomenaphthone | 112 | Hydrocortisone | 214d |
| Testosterone phenylpropionate | 115 | Hydrocortisone acetate | 220d |
| Calciferol | 116 | Prednisolone | 229d |
| Progesterone | 128 | Prednisone | 230d |
| Aspirin (Table 16) | 136 | Dienoestrol | 232 |
| Cholesterol | 148 | Phenolphthalein | 258 |

## Table P2. Compounds containing C, H, N, (and O)

| | B.p. | | M.p. |
|---|---|---|---|
| Amphetamine | 200 | Practolol | 144 |
| Nikethamide | 280 | Hexobarbitone | 145 |
| | | p-Aminosalicylic acid | 150 |
| | M.p. | Codeine | 156 |
| Amethocaine | 44 | Salbutamol | 156 |
| Lignocaine | 67 | Amylobarbitone | 157 |
| Hydroxyquinoline | 76 | Quinidine | 168 |
| Glutethimide | 86 | Paracetamol | 170 |
| Benzocaine (Tables 9 and 19) | 90 | Isoniazid | 172 |
| Oxyphenbutazone | 96 | Quinine | 174 |
| Pholocodine | 97 | Cyclobarbitone | 171 |
| Quinalbarbitone | 100 | Levallorphan tartrate | 176 |
| Mepyramine maleate | 101 | Phenobarbitone | 177 |
| Meprobamate | 104 | Barbitone | 190 |
| Phenylbutazone | 106 | Adrenaline | 212d |
| Amidopyrine | 107 | Nitrazepam | 229 |
| Phenazone | 111 | Mefenamic acid | 230 |
| Acetanilide (Table 7) | 114 | Nicotinic acid (Table 17) | 235 |
| Atropine | 114 | Caffeine | 236 |
| Levorphanol tartrate | 116 | Nitrofurantoin | 255 |
| Butobarbitone | 124 | Theophylline | 271 |
| Bisacodyl | 135 | Levodopa | 277d |
| Phenacetin (Table 7) | 135 | Primidone | 280 |
| Methoin | 138 | Methyldopa | 290 |
| Salicylamide (Tables 6 and 31) | 139 | Theobromine | 290 |
| Orthocaine | 143 | Phenytoin | 296d |

## Table P3. Compounds containing C, H, Halogen (and O)

| | B.p. | | M.p. |
|---|---|---|---|
| Halothane | 50 | Chloroxylenol (Table 31) | 115 |
| Ethchlorvynol | 174 | Chlorotrianisene | 118 |
| | | 4-Chloro-2-methylphenoxyacetic acid | 120 |
| | M.p. | Ethacrynic acid | 122 |
| Chlorocresol | 65 | 2,4-Dichlorophenoxyacetic acid | 138 |
| Chlorbutol | 77 | Fludrocortisone acetate | 225 |
| Butyl chloral hydrate | 78 | Betamethasone | 246d |
| Chlorphenesin | 81 | Dexamethasone | 255d |
| Dicophane | 109 | Fluoxymesterone | 278 |

## Table P4. Compounds containing C, H, N, halogen (and O)

| | M.p. | | M.p. |
|---|---|---|---|
| Chlorambucil | 67 | Amylocaine hydrochloride | 177 |
| Lignocaine hydrochloride | 77 | Pethidine hydrochloride | 189 |
| Cetylpyridinium chloride | 80 | Amitriptyline hydrochloride | 197 |
| Alprenolol hydrochloride | 111 | Morphine hydrochloride | 200d |
| Carbromal | 120 | Ephedrine hydrochloride | 218 |
| Flufenamic acid | 125 | Nortriptyline hydrochloride | 218 |
| Chlorpheniramine maleate | 133 | Procyclidine hydrochloride | 227 |
| Diazepam | 134 | Diamorphine hydrochloride | 229 |
| Chloramphenicol | 151 | Aminacrine hydrochloride | 234 |
| Procaine hydrochloride | 155 | Gallamine triethiodide | 235d |
| Indomethacin | 162 | Methadone hydrochloride | 236 |
| Procainamide hydrochloride | 167 | Antazoline hydrochloride | 240 |
| Imipramine hydrochloride | 170 | Pyrimethamine | 242 |
| Diphenhydramine hydrochloride | 170 | Decamethonium iodide | 246 |
| Isoprenaline hydrochloride | 172 | Nalorphine hydrobromide | 260d |
| Methylamphetamine hydrochloride | 173 | Acetrizoic acid | 280d |

## Table P5.  Compounds containing C, H, N, S (and O)

| | M.p. | | M.p. |
|---|---|---|---|
| Phenoxymethylpenicillin | 124 | Quinidine sulphate | 200d |
| Carbimazole | 125 | Sulphathiazole | 201 |
| Isoprenaline sulphate | 128d | Sulphadimethoxine | 204 |
| Tolbutamide | 128 | Orciprenaline sulphate | 205 |
| Poldine methylsulphate | 137 | Sulphamethizole | 211 |
| Sulphanilamide (Table 10) | 165 | Propylthiouracil | 220 |
| Dapsone | 176 | Saccharin (Table 6) | 230 |
| Sulphacetamide | 181 | Sulphamerazine | 235d |
| Succinylsulphathiazole | 189d | Procainamide sulphate | 236d |
| Sulphaguanidine | 191 | Sulphadiazine | 255d |
| Sulphapyridine | 191 | Acetazolamide | 258 |
| Atropine sulphate | 194 | Ephedrine sulphate | 258d |
| Sulphafurazole | 198 | Phthalylsulphathiazole | 272d |
| Sulphadimidine | 198 | Amphetamine sulphate | 300d |
| Probenecid | 199 | Mercaptopurine | 300d |

## Table P6.  Compounds containing C, H, N, S, Halogen (and O)

| | M.p. | | M.p. |
|---|---|---|---|
| Chlorpropamide | 128 | Promethazine hydrochloride | 223 |
| Promazine hydrochloride | 181 | Bendrofluazide | 225d |
| Chlorpromazine hydrochloride | 196 | Chlorothiazide | 343 |
| Chlorthalidone | 220 | | |

# INDEX